INTERACTING MACROMOLECULES

The Theory and Practice of Their Electrophoresis, Ultracentrifugation, and Chromatography

Molecular Biology

An International Series of Monographs and Textbooks

Editors

BERNARD HORECKER

Department of Molecular Biology
Albert Einstein College of Medicine
Yeshiva University
Bronx, New York

NATHAN O. KAPLAN

Department of Chemistry
University of California
At San Diego
La Jolla, California

JULIUS MARMUR

Department of Biochemistry
Albert Einstein College of Medicine
Yeshiva University
Bronx, New York

HAROLD A. SCHERAGA

Department of Chemistry
Cornell University
Ithaca, New York

HAROLD A. SCHERAGA. Protein Structure. 1961.

STUART A. RICE AND MITSURU NAGASAWA. Polyelectrolyte Solutions: A Theoretical Introduction, *with a contribution by Herbert Morawetz*. 1961

SIDNEY UDENFRIEND. Fluorescence Assay in Biology and Medicine. Volume I—1962. Volume II—1969

J. HERBERT TAYLOR (Editor). Molecular Genetics. Part I—1963. Part II—1967

ARTHUR VEIS. The Macromolecular Chemistry of Gelatin. 1964

M. JOLY. A Physico-chemical Approach to the Denaturation of Proteins. 1965

SYDNEY J. LEACH (Editor). Physical Principles and Techniques of Protein Chemistry. Part A—1969. Part B in preparation

KENDRIC C. SMITH AND PHILIP C. HANAWALT. Molecular Photobiology: Inactivation and Recovery. 1969

RONALD BENTLEY. Molecular Asymmetry in Biology. Volume I—1969. Volume II—1970

JACINTO STEINHARDT AND JACQUELINE A. REYNOLDS. Multiple Equilibria in Protein. 1969

DOUGLAS POLAND AND HAROLD A. SCHERAGA. Theory of Helix-Coil Transitions in Biopolymers. 1970

JOHN R. CANN. Interacting Macromolecules: The Theory and Practice of Their Electrophoresis, Ultracentrifugation, and Chromatography. 1970

(INTERACTING MACROMOLECULES)

The Theory and Practice of Their Electrophoresis, Ultracentrifugation, and Chromatography

JOHN R. CANN
DEPARTMENT OF BIOPHYSICS
UNIVERSITY OF COLORADO MEDICAL CENTER
DENVER, COLORADO

with a contribution by
WALTER B. GOAD
UNIVERSITY OF CALIFORNIA
LOS ALAMOS SCIENTIFIC LABORATORY
LOS ALAMOS, NEW MEXICO

1970

ACADEMIC PRESS   New York and London

COPYRIGHT © 1970, BY ACADEMIC PRESS, INC.
ALL RIGHTS RESERVED
NO PART OF THIS BOOK MAY BE REPRODUCED IN ANY FORM,
BY PHOTOSTAT, MICROFILM, RETRIEVAL SYSTEM, OR ANY
OTHER MEANS, WITHOUT WRITTEN PERMISSION FROM
THE PUBLISHERS.

ACADEMIC PRESS, INC.
111 Fifth Avenue, New York, New York 10003

United Kingdom Edition published by
ACADEMIC PRESS, INC. (LONDON) LTD.
Berkeley Square House, London W1X 6BA

LIBRARY OF CONGRESS CATALOG CARD NUMBER: 75-123362

PRINTED IN THE UNITED STATES OF AMERICA

CONTENTS

Preface — vii

Acknowledgments — ix

I. Principles of Electrophoresis and Ultracentrifugation — 1

Electrophoresis — 3
Ultracentrifugation — 19
Appendix I: Some Other Methods of Sedimentation Analysis — 40
Appendix II: Total Chemical Potential — 43
References — 44

II. Weak-Electrolyte Moving-Boundary Theory — 47

Macromolecular Isomerization — 48
Weak-Electrolyte Moving Boundary — 50
Interactions of Proteins with Small Ions — 58
Interactions between Different Macromolecules — 85
References — 91

III. Analytical Solution of Approximate Conservation Equations — 93

Analytical Solution of Approximate Conservation Equations — 93
The Similarity Transformation — 95
The Gilbert Theory — 102
The Gilbert–Jenkins Theory — 133
References — 149

IV. Numerical Solution of Exact Conservation Equations — 152

Factors Governing the Precise Shape of Reaction Boundaries — 152
Zone Transport of Reacting Systems — 171
Interaction of Macromolecules with Small Molecules — 176
References — 205

V. Numerical Methods — 207
Walter B. Goad

References — 217
Appendix: A Computer Code for Sedimentation Calculations — 217

VI. Practical Implications — 224

Velocity Sedimentation — 225
Electrophoresis — 228
Chromatography — 232
References — 233

Author Index — 235

Subject Index — 241

PREFACE

One usually associates methods of mass transport such as electrophoresis, ultracentrifugation, and gel filtration with the analysis and fractionation of biological tissues and fluids and the precise characterization of their highly purified components in terms of inherent state of homogeneity, size, shape, and electrical charge. But, another application is emerging as one of comparable importance, namely, detection and characterization of macromolecular interactions with respect to stoichiometry and thermodynamic and kinetic parameters. Although these capabilities were recognized many years ago by Tiselius[1] and by Longsworth and MacInnes,[2] developments in this area did not keep pace with purely analytical applications. This is attributable in part to the fact that, aside from elaboration of the weak electrolyte moving-boundary theory utilizing the concepts of constituent mobilities and concentrations, theoretical advances were for a long time impeded by mathematical difficulties. During the past decade or so substantial progress has been made in understanding the transport of interacting systems. These new insights stem from the realization by Gilbert[3,4] that, when obtainable, asymptotic analytical solutions of otherwise intractable transport equations are often good approximations to actual experimental situations. The theoretical treatments of Gilbert[3,4] and Gilbert and Jenkins[5] have had a profound influence on the way we think about the sedimentation and electrophoretic behavior of interacting systems, particularly with respect to the nature of reaction boundaries shown by self-associating macromolecules and by mixtures of different macromolecules which complex with one another. Thus, it was demonstrated for the first time that reaction boundaries may be bimodal even for rapidly established equilibria. Moreover, the areas and

---

[1] A. Tiselius, *Z. Physik. Chem.* **124,** 449 (1926).
[2] L. G. Longsworth and D. A. MacInnes, *J. Gen. Physiol.* **25,** 507 (1942).
[3] G. A. Gilbert, *Discussions Faraday Soc.* **20,** 68 (1955).
[4] G. A. Gilbert, *Proc. Roy. Soc.* (*London*), **A250,** 377 (1959).
[5] G. A. Gilbert and R. C. Ll. Jenkins, *Proc. Roy. Soc.* (*London*) **A253,** 420 (1959).

velocities of the two peaks in general do not bear a simple relationship to the concentrations and mobilities of the species in the initial equilibrium mixture. Like most advances of this sort, both theories pose new and interesting questions which have stimulated further theoretical investigations. By and large these have involved the use of high-speed electronic computers for obtaining accurate numerical solutions of transport equations which retain the second-order diffusional term. It thereby becomes possible to examine the effect of various molecular and environmental factors on the precise shape of reaction boundaries shown by a variety of macromolecular interactions. The latter range in complexity from the relatively simple case of rapid dimerization to cooperative macromolecule–small-molecule interactions. In addition to the predicative insight necessary for maximally effective application of the several methods of mass transport to interacting systems, such theory is providing the understanding required for unambiguous interpretation of their more common analytical applications.

Such is the subject of this monograph which carries the reader step by step from the principles of mass transport of noninteracting systems through formulation of the weak electrolyte moving-boundary theory and analytical solution of approximate transport equations for certain types of interactions to the most recent computer computations on ligand-mediated association–dissociation reactions. With respect to the latter, considerations set forth by Dr. Walter B. Goad lead to a code presented in FORTRAN language for accurate and rapid sedimentation calculations. Actual experimental systems have been chosen to illustrate the procedures and precautions required to assure accurate interpretation of sedimentation, electrophoretic, and other transport patterns in terms of the several parameters characterizing the reaction involved. Nor has sight been lost of the implications of macromolecular interactions for the more conventional analytcial and separatory applications of these methods. Thus, the development of the subject matter is designed to meet the needs not only of molecular biologists who wish to use transport methods for fundamental investigations on macromolecular interactions *per se* but also those for whom sedimentation, electrophoresis and chromatography are ancillary to their research. Finally, this monograph serves to introduce graduate students to an important and rapidly expanding area of molecular biology.

ACKNOWLEDGMENTS

The authors are grateful to their colleagues for many helpful discussions during preparation of this monograph and for criticisms of the manuscript or parts therof. We also wish to thank authors and copyright holders for their permission to reproduce figures from the literature. Preparation of this monograph was made possible in part by Grant-in-Aid 5R01 AI0482 from the National Institute of Allergy and Infectious Diseases, United States Public Health Service. This is Contribution Number 367 from the Deparment of Biophysics, University of Colorado Medical Center, Denver, Colorado.

INTERACTING MACROMOLECULES

The Theory and Practice of Their Electrophoresis, Ultracentrifugation, and Chromatography

1 / PRINCIPLES OF ELECTROPHORESIS AND ULTRACENTRIFUGATION

Methods of mass transport such as electrophoresis, ultracentrifugation, gel filtration, and chromatography are powerful means for separating macromolecules, and their importance in modern biology has steadily grown with the years. Classically, one considers these methods in terms of fractionation and analysis; in fact, they have found extensive application to analysis of biological tissues and fluids, characterization of different molecular forms of enzymes and other biologically important macromolecules and their subunits, and to genetic analysis. But another application is emerging as one of potentially comparable importance—namely, detection and characterization of macromolecular interactions. Its breadth and significance in providing a molecular understanding of biological processes are illustrated by studies on: the specific combination of two different proteins to give an active enzyme, as in the case of tryptophan synthetase from *Escherichia coli*; the subunit structure of hemoglobin and enzymes like lactic dehydrogenase; the specific combination of proteolytic enzymes with their macromolecular substrates and inhibitors and of antigens with their respective antibodies; complexing of *E. coli* repressor protein with *lac* operon DNA and of ribosomes with messenger RNA; and the allosteric interaction of aspartate transcarbamylase with its ligands.

The capabilities of ultracentrifugation and electrophoresis for detection and characterization of macromolecular interactions in terms of thermodynamic and kinetic parameters were recognized many years ago by Tiselius (*1*) and by Longsworth and MacInnes (*2*). However, developments in this area did not keep pace with purely analytical applications. This is attributable in part to the fact that, aside from elaboration of the weak-electrolyte moving-boundary theory utilizing the concepts of constitutive mobilities and concentrations, theoretical advances were for a long time impeded by mathematical difficulties; and predictive theoretical insight is required for maximally effective application of these methods to interacting systems. In addition, such theory provides the understanding required for unambiguous interpretation of the more common analytical applications. For example, it has often been assumed that zone electrophoresis is immune to the several nonideal effects which sometimes complicate moving-boundary electrophoresis; but recent theoretical and experimental developments make it clear that the same caution must be exercised in interpreting zone electrophoretic patterns as moving-boundary patterns.[1] In particular, cognizance must be taken of the fact that multiple zones need not necessarily indicate inherent heterogeneity.[2] Thus, a single macromolecule interacting reversibly with a small, uncharged constituent of the solvent can give two zones despite instantaneous establishment of equilibrium.

During the past decade or so, substantial progress has been made in understanding the transport of interacting systems. These advances were made possible by the realization that, when obtainable, asymptotic analytical solutions of otherwise intractable transport equations are often good approximations to actual experimental situations and, in part, by the application of high-speed electronic computers for obtaining accurate numerical solutions. In many instances, the theory has been formulated in terms of a particular transport method, but the predictions are equally valid for other methods. Consider, for example, the theory of sedimentation of reversibly polymerizing macromolecular systems in which monomer coexists with a single polymer. For trimerization or higher-order polymerizations, the ultracentrifuge patterns will show two sedimentating peaks despite instantaneous establishment of equilibrium. Likewise, the descending (but not ascending) moving-boundary electrophoretic patterns will show two peaks provided the electrophoretic mobility of the polymer is greater than that of the monomer. Both predictions have been realized experimentally for the low-temperature tetramerization of β-lactoglobulin A in acidic media.

Before proceeding with a detailed treatment of the mass transport of reversibly reacting macromolecules, this chapter will be devoted to a survey of the principles of electrophoresis and ultracentrifugation of nonreacting systems. This approach permits development of a theoretical framework which can be readily extended to include reacting molecules. In this regard, it seems

---

[1] This must be so because the moving-boundary pattern is simply a differential plot of the summation of a great many, slightly displaced zone patterns, at least in the limit of noninteracting systems. Even in the case of interacting systems, the differences are only quantitative.

[2] By inherent heterogeneity is meant a mixture of two or more stable noninteracting and nontautomeric macromolecules possessing distinctly different electrophoretic mobilities.

appropriate to conclude with a discussion of the conditions for sedimentation equilibrium of chemically reacting systems.

ELECTROPHORESIS

In its restricted sense, the term "electrophoresis" refers to the movement of charged colloidal particles and macromolecular ions under the influence of an electric field; but recent usage includes viruses, biological cells, and subcellular organelles, and small organic molecules such as amino acids. Depending upon the sign of their net charge, they migrate either to the cathode or anode. Differences in migration velocities provide a sensitive means of separating substances from their mixtures, which are otherwise difficult to fractionate. Indeed, electrophoresis is often the only method available for the quantitative analysis and fractionation of biological tissues and fluids and the characterization of their purified components.

The development of electrophoresis from a chance observation by the Russian physicist Reuss (3) in 1809 to its present status as a valuable analytical method illustrates the long evolutionary process through which a physical method must often pass before it can become a routine laboratory technique. Thus, moving-boundary electrophoresis was described by Lodge (4) in 1886 and developed to a high degree of perfection for the study of transference numbers of inorganic electrolytes by MacInnes and Longsworth (5) at the turn of the third decade of this century. The first application to the study of proteins was made in 1892 by Picten and Linder (6), who showed that hemoglobin is negatively charged in alkaline solution and positively charged in acid solution. But it was not possible to make precise measurements on purified proteins and complex biological materials until Tiselius (7) described his adaptation of the moving-boundary method in 1937. Since then, however, progress has been rapid. The abundant results obtained with the Tiselius apparatus as modified by Longsworth (8) have contributed significantly to the remarkable strides made in the past two decades in the various branches of biochemistry and medicine. With the unequivocal demonstration (9–14) in 1950 of the value of paper electrophoresis for the separation of proteins, zone electrophoresis on solid supports and in gels gained tremendous popularity. It is the method of choice in situations where separation rather than precise physicochemical measurements is the overriding consideration. Zonal methods are suitable for the analysis of very small quantities of material by fairly simple procedures, which permit easy visualization of zones and isolation of fractions. Moreover, they are capable of extremely high resolving power, made possible by combination of electromigration with molecular sieving in starch gel and polyacrylamide gels. Also, zone electrophoresis can be used for the investigation of low-molecular-weight substances, which are difficult to analyze by the moving-boundary method. These advantages are gained, however, at the sacrifice of accuracy and precision, particularly with respect to the determination of mobilities from migration rates in solid supports. This loss of accuracy is a serious consideration if one wishes to use zone electrophoresis for identification and characterization of macromolecules and other charged substances in the same way as in moving-boundary electrophoresis. In any case, it is evident that electrophoresis continues to play an indispensable role in the development of many areas of biology and medicine.

Theory of Electromigration. The velocity of a particle in free solution when the electric field acting on it is 1 V cm^{-1} is called its electrophoretic mobility. The dimensions of mobility are cm^2 sec^{-1} V^{-1}, and its sign is the same as that of the net electrical charge on the particle. Although in a given medium the mobility is a characteristic property of the particle, it generally varies with the composition of the solution. This behavior is especially true of protein ions, where the mobility depends upon the pH and ionic strength of the solution, and, in many cases at least, upon the nature of the solvent electrolyte. An understanding of this behavior ultimately depends upon the establishment of a theoretical relationship between the mobility and various molecular parameters such as electrical charge and frictional coefficient, and the elucidation of the dependence of these parameters upon the composition of the solvent medium. Theoretical interpretation of the electrophoretic mobility has been considered in detail by several authors (*15–18*), and only a brief survey of the problem will be presented here.

Consider an isolated particle of charge Q and mass m suspended in a perfect insulator. A uniform electric field of intensity E will exert a force QE on the particle. Upon accelerating from rest, it experiences an opposing frictional force $f\,dx/dt$, where f is the frictional coefficient of the particle and dx/dt its instantaneous velocity. The resultant of these two forces is, by Newton's second law, equal to $m\,d^2x/dt^2$, so that the equation of motion is

$$m\,d^2x/dt^2 = QE - f\,dx/dt \tag{1}$$

The solution of Eq. (1) is

$$dx/dt = (QE/f)[1 - e^{-(f/m)t}] \tag{2}$$

But, since the value of $f/m \gg 1$ is of the order 10^{12}–10^{14} sec^{-1} for particles of molecular size in solution, the transient $e^{-(f/m)t}$ decays to $1/e$ in 10^{-14} to 10^{-12} sec, so that for times greater than about 10^{-11} sec, Eq. (2) reduces effectively to

$$dx/dt = QE/f \tag{3}$$

In other words, upon application of the electric field, the particle accelerates very rapidly until the electrical force is exactly balanced by the frictional force, after which time, it moves at the constant velocity given by Eq. (3). The frictional coefficient may be obtained from the diffusion coefficient D by means of the Einstein relationship

$$D = kT/f \qquad (4)$$

where k is Boltzmann's constant and T is the absolute temperature. The electrophoretic mobility μ for a particle of arbitrary shape is then given by the relation

$$\mu = \frac{1}{E}\frac{dx}{dt} = \frac{Q}{f} = \frac{QD}{kT} \qquad (5)$$

If the particle is a small sphere of radius a, the frictional coefficient is given by Stoke's law as $6\pi\eta a$, where η is the coefficient of viscosity of the solvent medium. The radius of the particle may be eliminated by introducing the potential at the surface of the sphere ψ_0, which is equal to $Q/\mathbf{D}a$, where $\mathbf{D}$ is the dielectric constant. The mobility of the sphere is then given by

$$\mu = \mathbf{D}\psi_0/6\pi\eta \qquad (6)$$

Equations (5) and (6) are developed for charged particles in a perfect insulator and must be modified for application to electrophoretic experiments on macromolecular ions in electrolyte solutions. The additional fundamental concept required for understanding electromigration in electrolyte solutions is that of the ionic atmosphere. As a consequence of the electrostatic force between the charge on the particle and the ions of the electrolyte, there are, on the average, more ions of unlike than of like sign in the neighborhood of the particle. In other words, the particle may be regarded as surrounded by an ionic atmosphere of opposite charge. The presence of the ionic atmosphere results in electrophoretic mobilities which are smaller than those predicted by Eqs. (5) and (6). Three factors contribute to this difference. First the ionic atmosphere lowers the value of the potential at the surface of the particle. One may visualize this effect by considering that the ionic atmosphere shields the charge on the particle from the applied electric field; that is, it decreases the effective charge, and thus the migration velocity. The second effect arises from the fact that the applied field also acts upon the ions of the ionic atmosphere. Since the charge of the ion cloud is opposite in sign to the charge of the particle, the force exerted by the electric field on the ion cloud tends to move it in a direction opposite to that of the particle, thus decreasing the migration velocity of the particle. This effect is referred to as the *electrophoretic effect* or *electrophoretic friction*, since it effectively increases the frictional force acting on the particle. Finally, the electric current continually carries new ions to and from the environment of the particle. This exchange of ions distorts the otherwise spherically symmetrical ion atmosphere because the ions approaching the particle require a finite time before their distribution can adjust itself to the field distribution near the particle. Similarly, the ions leaving the atmosphere cannot instantaneously assume a random distribution. As a result, the ionic atmosphere will trail behind the moving particle, thereby producing an electrostatic retarding force which decreases the velocity of the particle. The production of an asymmetrical ionic atmosphere is referred to as the *relaxation effect*. All three of these effects increase with increasing ionic strength of the solvent medium. Thus, other things remaining constant, the electrophoretic mobility is expected to decrease when the ionic strength is increased.

Henry (*18a*) considered the electromigration of a nonconducting particle of arbitrary radius in an electrolytically conductive medium. Assuming that the relaxation effect is negligible, he found the electrophoretic mobility to be given by the relation

$$\mu = \frac{\mathbf{D}}{4\pi\eta}\left[\zeta + 5a^3 \int_\infty^a \frac{\psi(r)}{r^6}\,dr - 2a^3 \int_\infty^a \frac{\psi(r)}{r^4}\,dr\right] \qquad (7)$$

where $\psi(r)$ is the potential in the ionic atmosphere, and ζ is the zeta potential, i.e., the potential at the surface of shear. A few words of explanation with regard to the zeta potential are in order. A layer of water may adhere so firmly to the particle that it cannot be set into motion either by an applied electric field or by motion of the liquid. Thus, the adhering layer of water must be considered as forming a part of the particle, and the potential determining the rate of electromigration is then taken as the potential at the boundary of the fixed and free liquid, i.e., the surface of shear. The zeta potential is determined by the charge inside this surface. This charge is not necessarily identical with the net charge of the particle, since some of the ions of the ionic atmosphere may be present within the surface of shear, thereby reducing the electrophoretic charge below that which would be determined analytically, as, for example, by acid–base titration.

On the assumption that the zeta potential is small and that the interionic attraction theory of Debye and Hückel is applicable,[3] Eq. (7) reduces to

$$\mu = \frac{\mathbf{D}\zeta}{6\pi\eta}\,g(\mathscr{H}a)$$

$$g(\mathscr{H}a) = 1 + \frac{\mathscr{H}^2 a^2}{16} - \frac{5\mathscr{H}^3 a^3}{48} - \frac{\mathscr{H}^4 a^4}{96} + \frac{\mathscr{H}^5 a^5}{96}$$

$$- \left(\frac{\mathscr{H}^4 a^4}{8} - \frac{\mathscr{H}^6 a^6}{96}\right)(\exp \mathscr{H}a) \int_\infty^a \frac{e^{-t}}{t}\,dt$$

---

[3] The use of the Debye–Hückel theory introduces an inconsistency into the development of the theory of electromigration insofar as it involves the assumption that the zeta potential is equal to the potential at the surface of the particle. However, this may not be too serious at very low ionic strengths, where only a negligible part of the charge in the double layer might be included within the surface of shear.

$$\mathcal{H} = \left(\frac{8\pi N\varepsilon^2}{1000\mathbf{D}kT}\right)^{1/2} \sqrt{\Gamma} \qquad (8)$$

where ε is the elementary charge; N is Avogadro's number; Γ is the ionic strength ($\Gamma = \frac{1}{2}\Sigma^i z_i^2 C_i$, where z_i is the charge of the ith ion, whose molar concentration is C_i); and $1/\mathcal{H}$ is the "thickness" of the ionic atmosphere. Values of $6/g(\mathcal{H}a)$ for various values of $\mathcal{H}a$ are presented in Table I.

TABLE I

HENRY'S FACTOR FOR SPHERES

$\mathcal{H}a$	$6/[g(\mathcal{H}a)]$	$\mathcal{H}a$	$6/[g(\mathcal{H}a)]$
0	6.000	5	5.173
1	5.844	10	4.843
2	5.631	25	4.38
3	5.450	100	4.11
4	5.298	∞	4.00

As already mentioned, Henry's treatment neglects consideration of the relaxation effect; more recent theoretical investigations (17, 18) indicate that, for monovalent supporting electrolytes and for zeta potentials in the range expected for proteins, the relaxation effect is, in fact, small. This justifies many applications of Eqs. (8) to the evaluation of zeta potentials from electrophoretic mobilities of proteins.[4] In most investigations, however, the net charge on the particle, rather than its zeta potential, is of primary interest. Evaluation of the charge is more difficult than evaluation of the zeta potential. Some of the difficulties have already been implied. For small zeta potentials, the charge is usually calculated from the Debye–Hückel relationship

$$Q = \mathbf{D}\zeta a(1 + \mathcal{H}a) \qquad (9)$$

If the potential is large, evaluation of Q is very difficult (17) except for flat surfaces, that is, for very large particles approximately spherical or oblate elliptical in shape and possessing a smooth surface, such as, for example, in the case of mammalian erythrocytes. The appropriate equations are given by

---

[4] A recent theory (18) including this effect shows that, for monovalent supporting electrolytes, the relaxation effect is negligible when $\mathcal{H}a$ is very small or very large, but it is appreciable for moderate values of $\mathcal{H}a$, and increases sharply with increasing ζ. Overbeek and Wiersema (18) indicate how one can decide whether Henry's equation, Eq. (8), is appropriate for computation of ζ from experimental values of μ. If it is not, recourse can be had to numerical solutions of a set of theoretical equations which do include the relaxation effect.

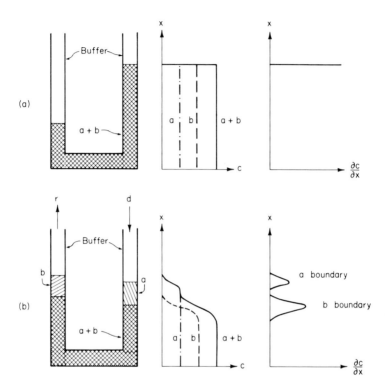

FIG. 1. Resolution of a mixture of two electrophoretic components, a and b, under ideal conditions. Diagrammatic representation of the boundaries in the U-shaped moving-boundary electrophoresis cell and plots of concentration c and concentration gradients, $\partial c/\partial x$ as functions of height x in the descending limb of the cell: (a) initial conditions, (b) conditions after application of electric field for a time sufficient to resolve completely the moving boundaries corresponding to components a and b. Original mixture contains one part of a and two parts of b. Mobility of b is greater than that of a. [From Cann (23).]

Abramson et al. (15). These authors also give equations for computing Q from the electrophoretic mobilities of disk- or rod-shaped particles of low potential.

Recently (19), the equation for rods has been applied to deoxyribonucleic acid (DNA). The mobilities at various NaCl concentrations (0.01 M Tris buffer, pH 7.5) were those obtained in the elegant experiments of Olivera et al. (20). The computed values of Q were used, in turn, to calculate equilibrium distributions of chloride ion across a membrane impermeable to DNA. Distributions computed in this manner from electrophoretic mobilities and those obtained experimentally from Donnan equilibrium measurements agreed to

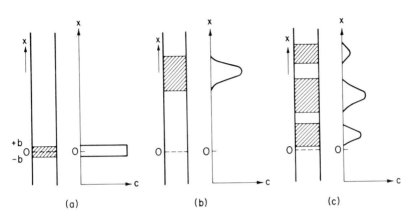

FIG. 2. Diagrammatic representation of zone electrophoresis on a supporting medium: (a) initial conditions; (b) the single moving zone shown by an electrophoretically homogeneous substance; (c) separation of a mixture of three substances with distinctly different rates of migration. [From Cann (23).]

within 2% over a twentyfold range of NaCl concentration. In essence, the electrophoretic charge of DNA agrees well with its net charge as derived from a thermodynamic measurement. Similarly, when the electrophoretic charge calculated for ovalbumin assuming spherical shape [Eqs. (8) and (9)] is compared with the charge obtained from membrane-potential measurements in the same phosphate buffer, pH 7.10, the agreement is very satisfactory over a wide range of ionic strengths (21). However, these results must be viewed with caution, for it is not clear whether one should expect agreement between electric potentials calculated from electrokinetic data and those derived from thermodynamic measurements (18). Also, the calculations for DNA make somewhat free use of the Debye–Hückel theory. Nevertheless, it would seem that the theory of electromigration outlined above is a good approximation to the problem of relating electrophoretic mobility to the molecular parameters of net charge and frictional coefficient and to ionic strength. It has also proved a valuable guide to understanding the dependence of mobility on solvent composition. Thus, the type of buffer electrolyte is important, since the net charge is determined not only by the state of ionization of the acidic and basic groups on the protein, which is a function of pH, but also by the binding of buffer ions other than H^+ by the protein molecule. In a number of instances, deviations of electrophoretic charge from titration charge have been attributed to the binding of salt ions; in fact, electrophoresis is a useful tool for studying macromolecule–small-ion interactions. Binding of H^+ and other

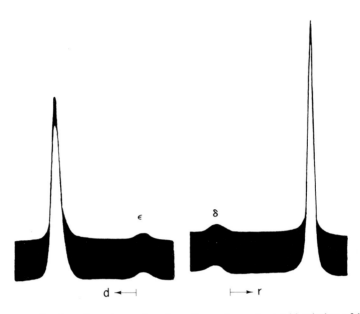

FIG. 3. Moving-boundary electrophoretic patterns shown by a 1% solution of bovine serum albumin in 0.1 ionic strength sodium diethylbarbiturate buffer at pH 8.6. By convention, the patterns have been rotated 90° from the position in which they were recorded. The direction of migration is shown by the arrows labeled to indicate whether the boundary is ascending (r) or descending (d). The vertical line at the tail of the arrow indicates the position of the initial boundary. [From Cann (23).]

small ions can also affect the frictional coefficient through changes in either conformation or state of aggregation. The combined use of electrophoresis and a hydrodynamic method such as ultracentrifugation or viscosity measurements may sometimes be indicated in order to evaluate the relative importance of changes in net charge and frictional coefficient with changes in electrolyte environment. Finally, the theory predicts that the mobility will decrease with increasing ionic strength. Although this is generally found to be the case, occasionally a protein, such as aldolase in phosphate buffer, will exhibit just the opposite behavior. Here also, deviations are usually interpreted in terms of binding of buffer ions other than H^+; but in some instances other factors, such as changes in frictional coefficient, could conceivably be important. In any event, when quoting a value for electrophoretic mobility, it is imperative to state the conditions of pH, ionic strength, and buffer composition.

Phenomenological Theory. In moving-boundary electrophoresis, an infinitely sharp initial boundary is formed between the dialyzed solution of macro-

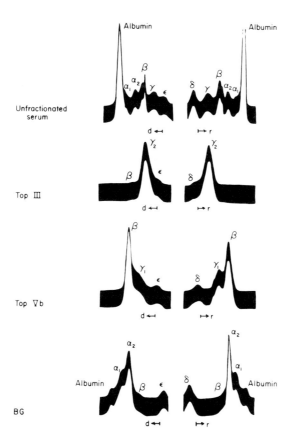

Fig. 4. Moving-boundary electrophoretic patterns of human serum and some of its fractions in 0.1 ionic strength sodium diethylbarbiturate buffer at pH 8.6. The fractions were obtained by the method of electrophoresis convection. [From Cann and Loveless (29).]

molecular ions and the buffer solution in a vertical electrophoresis column of rectangular cross section, with the denser solution underneath. On the other hand, in zone electrophoresis, a narrow zone of the solution of macromolecule is applied to a strip, slab, column, or fiber of supporting material containing the buffer. Upon application of the electric field, the boundary or zone is caused to move along the electrophoresis column. An electrophoretically homogeneous macromolecule will show a single moving boundary or zone; but with a heterogeneous specimen, the initial boundary or zone will resolve into two or more moving boundaries or zones, each corresponding to one of the electrophoretically distinct components of the original mixture (Figs. 1

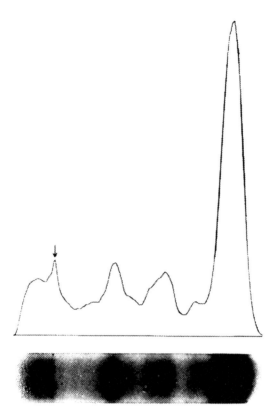

Fig. 5. Paper electrophoretic pattern of human serum in 0.05 ionic strength sodium diethylbarbiturate buffer, pH 8.6; dyed paper strip and optical scan of dyed strip. The anode is to the right, and the vertical arrow indicates the position of the initial zone. The major zone to the far right is albumin, the zone to the far left is γ-globulin. [From Block et al. (30).]

and 2). Records of the moving boundaries or zones obtained after some arbitrary time of electrophoresis are called electrophoretic patterns. Moving-boundary electrophoretic patterns are plots of concentration gradient (or refractive-index gradient, which is proportional to the concentration gradient) versus position in the electrophoresis column; zone patterns are plots of concentration versus position. Typical moving-boundary and zone patterns are presented in Figs. 3–6. The theoretical problem to which we address ourselves is the computation of the electrophoretic patterns and the rates of movement and changes in shapes of the boundaries or zones as they develop with time. One proceeds by considering first the mass transport of an electrophoretically

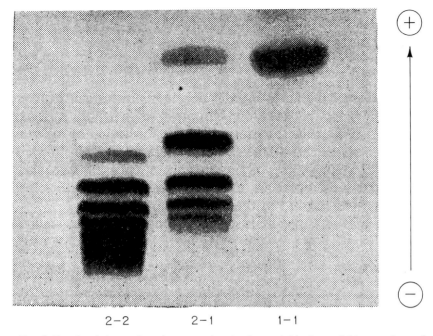

FIG. 6. Starch gel electrophoresis pattern showing haptoglobin–hemoglobin complexes of the three phenotypes, 2–2, 2–1, and 1–1, stained with a benzidine reagent. [From Harris (*31*).]

homogeneous macromolecular ion and then extending these considerations to mixtures of different macromolecules.

If no field is applied, the initially sharp boundary or zone will spread with time owing to free diffusion of the macromolecule. During electrophoresis, however, free diffusion is modified by the presence of the external electric field, which produces a directed flow of the macromolecular ions. The process of modifying free diffusion by an external force is called forced diffusion. The fundamental differential equation of forced diffusion, often referred to as the conservation equation, is simply a statement of the conservation of mass during transport. Its solution gives the electrophoretic pattern. The differential equation is readily derived by imposing material balance while following the motion of charged macromolecules during a short time interval as they pass into and out of a thin cross-sectional slice of the electrophoresis column (Fig. 7) under the influence of the electric field and random thermal motion.

The amount of macromolecule entering the slide across its face at position x in the differential time interval dt due to the random thermal motion of translational diffusion is given by Fick's first law as

$$-D(\partial c/\partial x) \mathscr{A} \, dt \tag{10}$$

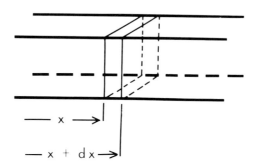

FIG. 7. Thin cross-sectional slice of electrophoresis column. Cross-sectional area is $\mathscr{A}$; thickness of slice is dx; electromigration in the x direction is indicated by arrows.

where $\partial c/\partial x$ is the concentration gradient at that face, and D is the diffusion coefficient of the macromolecule. Here, c denotes its concentration in either mole per cubic centimeter or grams per cubic centimeter. During the same time interval, the amount of macromolecule

$$-\left(D+\frac{\partial D}{\partial x}dx\right)\left(\frac{\partial c}{\partial x}+\frac{\partial^2 c}{\partial x^2}dx\right)\mathscr{A}\,dt \qquad (11)$$

leaves the slice by diffusion across the face at position $x + dx$. The amount of material accumulated in the slice due to diffusion is given by the difference between expressions (10) and (11),

$$-D\frac{\partial c}{\partial x}\mathscr{A}\,dt+\left(D+\frac{\partial D}{\partial x}dx\right)\left(\frac{\partial c}{\partial x}+\frac{\partial^2 c}{\partial x^2}dx\right)\mathscr{A}\,dt \qquad (12)$$

which, if one ignores the cross term containing a differential of higher order, reduces to

$$\frac{\partial}{\partial x}\left(D\frac{\partial c}{\partial x}\right)\mathscr{A}\,dx\,dt \qquad (13)$$

In passing, it should be noted that expression (13) is a statement of Fick's second law of diffusion.

The amount of macromolecule simultaneously entering the slice through the face at x due to electromigration is

$$\mu E c \mathscr{A}\,dt \qquad (14)$$

where E is the electric field strength and μ is the electrophoretic mobility. The amount leaving through the face at $x + dx$ is

$$\left(\mu + \frac{\partial \mu}{\partial x} dx\right)\left(E + \frac{\partial E}{\partial x} dx\right)\left(c + \frac{\partial c}{\partial x} dx\right)\mathscr{A} dt = \left[\mu E c + \frac{\partial(\mu E c)}{\partial x} dx\right]\mathscr{A} dt \quad (15)$$

The accumulation in the slice due to electromigration is the difference between expressions (14) and (15),

$$\mu E c \mathscr{A} dt - \left[\mu E c + \frac{\partial(\mu E c)}{\partial x} dx\right]\mathscr{A} dt = -\frac{\partial(\mu E c)}{\partial x}\mathscr{A} dx\, dt \quad (16)$$

The total accumulation of macromolecule in the slice due to diffusion and electromigration, $(\partial c/\partial t)\mathscr{A} dx\, dt$, is equal to the sum of expressions (13) and (16),

$$\frac{\partial c}{\partial t}\mathscr{A} dx\, dt = \frac{\partial}{\partial x}\left(D \frac{\partial c}{\partial x}\right)\mathscr{A} dx\, dt - \frac{\partial(\mu E c)}{\partial x}\mathscr{A} dx\, dt \quad (17)$$

or

$$\frac{\partial c}{\partial t} = \frac{\partial}{\partial x}\left(D \frac{\partial c}{\partial x}\right) - \frac{\partial(\mu E c)}{\partial x} \quad (18)$$

Simplification of this equation is afforded by assuming (1) D and μ independent of c, and (2) ideal electrophoresis in which none of the electric current is carried by the macromolecular ion, which simply drifts in the field maintained by the current carried by the ions of the solvent medium. In the latter event, gradients of conductance and pH will not be established in the electrophoresis column by the electrophoretic process, so that E and μ may be considered as constants. Given these approximations, Eq. (18) reduces to

$$\frac{\partial c}{\partial t} = D \frac{\partial^2 c}{\partial x^2} - \mu E \frac{\partial c}{\partial x} \quad (19)$$

which is the forced-diffusion equation for ideal electrophoresis.

The implications of the term "forced diffusion" are seen with greatest clarity upon transformation of Eq. (19) to the moving coordinate system, $y = x - \mu E t$,

$$\frac{\partial c}{\partial t} = D \frac{\partial^2 c}{\partial y^2} \quad (20)$$

This transformed equation is Fick's second law for diffusion in the moving coordinate system. In other words, an electrophoretic moving boundary or zone is simply a diffusion boundary or zone being transported along the electrophoresis column at the constant velocity μE due to the action of an external electrical force on the charged solute molecules under examination. Realization of this fact is a powerful mathematical lever, since coordinate transformations of known solutions of Fick's second law for diffusion from boundaries or zones yield directly solutions of the differential equation for ideal electrophoresis. These latter solutions define the shape and the position of ideal electrophoretic boundaries and zones as a function of time of electrophoresis.

Thus, the transformed solution for moving-boundary electrophoresis,

$$\frac{\partial c}{\partial x} = \frac{c_0}{2(\pi Dt)^{1/2}} \exp -\frac{(x-\mu Et)^2}{4Dt} \qquad (21)$$

shows that (1) the ideal moving boundary has the form of a Gaussian distribution; (2) the mobility is given by $\mu = \bar{x}/Et$, where the first moment $\bar{x}$ is taken about the position of the initial boundary; (3) the area under the gradient curve is equal to the concentration of the macromolecule, c_0, which is said to disappear across the boundary; and (4) the ascending and descending patterns are mirror images. The electrophoretic patterns of a solution of macromolecular ions containing not one, but two or more different noninteracting species with different electrophoretic mobilities will show a corresponding number of ideal moving boundaries (Fig. 1). The area under each boundary gives the concentration of the corresponding macromolecular species, and the rate at which the first moment of each peak moves is a measure of the mobility of that component. (Actual patterns, Figs. 3 and 4, are photographic records of refractive-index gradients versus position, so that areas give relative concentrations.) In practice, the conditions required for ideality are rarely completely realized (22, 23). As a consequence, (1) the ascending and descending patterns are usually somewhat nonenantiographic; (2) the ascending and descending patterns each show a small stationary boundary, the ε and δ boundaries, respectively, which can be ignored as far as determination of apparent composition is concerned; (3) apparent electrophoretic compositions determined by area measurements on either the ascending or descending pattern will differ from the actual composition; and (4) migration velocities of ascending boundaries are generally greater than those of the corresponding descending ones, and calculation of mobilities from ascending patterns is complicated. Deviations between apparent and actual compositions can be minimized by analysis of descending rather than ascending patterns. Extrapolation of apparent compositions determined at different total protein concentrations to zero protein concentration gives approximately the correct composition. The electrophoretic mobility of a protein showing a single moving boundary may be calculated directly from the displacement of the descending boundary and the conductance of the protein solution. With mixtures, however, this method gives the correct mobility only for the most rapidly moving component, the displacements of all the other boundaries giving values that are smaller than the corresponding mobilities. Consequently, the pH–mobility curve, and hence the isoelectric point of a given component of a mixture, should be determined, if possible, upon material which has been separated from the other components.

Departures from ideal behavior arise from the fact that in reality the protein does not simply drift in an electric field maintained exclusively by the current carried by the small ions of the supporting electrolytes. The protein itself actually carries a significant portion of the current. Consequently, if mass is to be conserved, the concentrations of the several ionic species must adjust across each boundary. For example, during resolution of the descending pattern, the concentrations of the small ions adjust across the initial boundary, thereby giving rise to the ε boundary. Likewise, the concentrations of both small ions and protein ions adjust across the initial ascending boundary to form the δ boundary. Thus, we see that the compositions of the solutions generated between the ε or δ boundary and the slowest-migrating descending or ascending moving boundary, respectively, differ from the buffer or the dialyzed protein solution used to form the initial boundary. The same pertains to all other solutions generated by the electrophoretic process, e.g., the solution between the a and b boundaries in Fig. 1. As a result, the conductance, and thus the electric field, change across each boundary. Since a given moving boundary migrates at the same rate as the corresponding protein molecules in the solution ahead of or behind the boundary, only the fastest descending boundary gives a correct mobility when the conductance of the dialyzed protein solution is used in the computations. Moreover, the area under a given boundary in a conventional electrophoretic pattern recorded using Schlieren optics is proportional to the total change in refractive index across that boundary. Since the concentrations of several ionic species change across the boundary, its area is a composite and not solely a measure of the concentration of the corresponding protein. For this reason, apparent electrophoretic composition departs from actual composition.

It is of considerable importance for certain types of experiments (*23a, 23b*) that the compositions of the various solutions generated by the electrophoretic process can be estimated (*22, 23*) from those of the buffer and dialyzed protein solution using the Dole strong-electrolyte moving-boundary theory (*23c*). Application of a theory of strong electrolytes to proteins, which are weak electrolytes, is justified by the fact that the variation in pH and ionic strength along the electrophoresis column is sufficiently small that the mobility of the protein is virtually constant.

The solution of Eq. (20) for zone electrophoresis is

$$c = \frac{c_0}{\sqrt{\pi}} \left[\int_0^{(b-y)/2\sqrt{Dt}} \exp(-z^2)\,dz - \int_0^{-(b+y)/2\sqrt{Dt}} \exp(-z^2)\,dz \right] \quad (22)$$

where b is the half-width of the initial zone. For a mixture of different species with different mobilities, the initial zone should resolve into a corresponding number of zones, each exhibiting this ideal shape (Fig. 2). The area under a given zone, when multiplied by the product of the width and thickness of the zone, gives the quantity of the corresponding component. For this highly idealized treatment only, the rates of zone migration are direct measures of electrophoretic mobility. In practice, deviations from ideality are tremendous. Although the dispersion of isolated zones can generally be fitted by probability integral distributions, at least in agar gels, there is considerably greater spreading of the zones during migration than predicted by Eq. (22). Departures of the rates of migration from ideality are often very marked indeed. For example, during paper electrophoresis of serum at pH 8.6, the γ-globulin zone migrates toward the cathode despite the fact that the γ-globulin molecule is negatively charged at this pH (Fig. 5). The migration is, of course, anodic with respect to the interstitial fluid which is flowing toward the cathode due to electroosmosis. Other factors which influence the rate of electromigration on supports include (1) the purely geometrical effect of the structure of the medium, which makes the charged particles and electric current travel through tortuous channels, as in the case of paper; (2) impedance of migration of macromolecular ions by mechanical resistance, as in silica gel; (3) molecular sieving effects, as in starch and polyacrylamide gels; (4) adsorption of the macromolecule to the supporting medium, which can cause trailing or even complete immobilization of a zone; and (5) changes in conductance and pH of the buffer in the supporting medium due to evaporation and other poorly understood effects. Nevertheless, by careful evaluation and control of these factors in paper, Kunkel and Tiselius (*24*) and Waldmann-Meyer (*25–28*) have derived values of electrophoretic mobilities and isoelectric points of several highly purified proteins which agree satisfactorily with those determined by the moving-boundary method. However, these mobility calculations necessitate certain approximations whose generality remains to be established.

In contrast to electrophoresis on solid supports and in gels, zone electrophoresis in a density gradient permits direct evaluation of electrophoretic mobilities simply by viscosity correction of migration rates per unit field. Indeed, Olivera and co-workers (*20*) have recently shown that, in the particular case of DNA, more accurate mobilities can be obtained by zone electrophoresis in a gradient of sucrose or D_2O than by moving-boundary electrophoresis.

Finally, it should be borne in mind that the various nonideal effects encountered in zone electrophoresis may not always be disadvantageous. Mention has already been made of the high resolving power achieved by combination of electromigration with molecular sieving in starch and polyacrylamide gels. Likewise, differences in extent of reversible adsorption to the supporting medium of different migrating species possessing the same electrophoretic mobility may result in their separation, thereby revealing an inhomogeneity which might otherwise go undetected. The severe spreading of zones of low-molecular-weight substances like peptides and amino acids during their separation under conventional conditions of low field strength (10–20 V cm^{-1}) led to the development of high-voltage electrophoresis—fields as high as 100 V cm^{-1}. At such high field strengths, the time required for separation is sufficiently short that diffusion is minimal, and sharp discrete spots or zones are obtained. Moreover, high-voltage electrophoresis in one dimension can be combined with chromatography in a second dimension, thereby increasing the resolving power of the method still further. Zone electrophoresis in gels has also been combined sequentially with immunodiffusion. Immunoelectrophoresis is perhaps the most sensitive means available for detection and identification of heterogeneity in purified proteins.

ULTRACENTRIFUGATION

Ultracentrifugation refers to the sedimentation of macromolecules under the influence of an intense centrifugal force. Differences in rates of sedimentation provide another powerful method for analysis and separation of biological materials, and precise determination of sedimentation rates allows characterization of highly purified macromolecules with respect to size, shape, and density. Many important results (32) were obtained in the years following construction of the first oil-turbine analytical ultracentrifuge by Svedburg and his co-workers in 1925–1926. The most important of these was the demonstration that native proteins consist either of a single or of a few molecular species each possessing a well-defined molecular mass and shape. Any remaining doubt that proteins are definite chemical substances possessing a unique structure was dispelled when Sanger (33) established the chemical structure of insulin, and Kendrew (34) established the three-dimensional X-ray crystallographic structure of myoglobin. It is also highly interesting that Svedberg (32) anticipated modern concepts of the subunit structure of proteins in his early formulation of the now defunct hypothesis of the multiple law for the molecular weights of proteins. Since the advent about 15 years ago of commercially available, electrically driven instruments, progress in the practice and theory of ultracentrifugation has been dramatic (35–41). Like electrophoresis, ultracentrifugation has become an indispensable research tool of the biochemist, molecular biologist, and clinical investigator. It is practiced either as analytical or preparative centrifugation, depending upon whether analysis is carried out during sedimentation or on fractions removed from the centrifuge tube after cessation of centrifugation. The principles of preparative centrifugation (37,

38, and Appendix I to this chapter) are essentially the same as for analytical centrifugation, and we shall focus our attention on the latter.

Basically, the analytical ultracentrifuge consists of a rotor which rotates about an axis at a constant velocity. Placed in the rotor, some distance from the axis of rotation, is a small cell containing the solution to be examined. The centrifugal force acting on the solution is directed radially outwards. Accordingly, the cell has the shape of a sector with its apex at the center of rotation in order to allow unobstructed sedimentation of the molecules along radii. The instrument is fitted with optical systems designed to record photographically either concentration or concentration gradient (more properly, refractive-index gradient) of the sedimenting molecules as a function of position in the cell. These records, which are called sedimentation patterns, may be obtained at such times during the sedimentation process as dictated by the nature of the particular experiment. The analytical ultracentrifuge lends itself to several independent methods (35, 36, 39–41, and Appendix I to this chapter) for the analysis of mixtures and the identification and characterization of separated components. Here, however, we shall be concerned only with the classical methods of sedimentation velocity and sedimentation equilibrium. A brief description of these methods follows. The reader is referred to Svedberg and Pedersen (32) and Schachman (35) for details.

By way of introduction, let us consider the forces acting on a molecule which is sedimenting through a solution of density ρ in an ultracentrifuge cell rotating with constant angular velocity ω rad sec^{-1}. According to Archimedes' principle, the buoyant mass of the molecule is $m = (M/N)(1 - \bar{v}\rho)$, where M is its molecular weight and $\bar{v}$ its partial specific volume. A centrifugal field of intensity $\omega^2 r$ will exert a force $m\omega^2 r$ on a molecule situated at a distance r from the center of rotation. When the centrifugal force is applied, the molecule accelerates very rapidly from rest, whereupon it experiences an opposing frictional force $f\, dr/dt$. In an extremely short time, on the order of 10^{-10} sec, the centrifugal force is exactly balanced by the frictional force, after which, the molecule moves at a velocity given by

$$m\omega^2 r = f\, dr/dt \tag{23}$$

The velocity per unit centrifugal field is called the sedimentation coefficient s of the molecule. The dimension of s is seconds, although sedimentation coefficients are reported in Svedbergs (S), where $1\,S = 10^{-13}$ sec. The relationship of the sedimentation coefficient to other molecular parameters is obtained by rearrangement of Eq. (23)

$$s = \frac{1}{\omega^2 r}\frac{dr}{dt} = \frac{m}{f} \tag{24}$$

It is instructive to compare this equation with Eq. (5) for the electrophoretic mobility. Whereas the electrophoretic mobility is equal to the ratio of net electrical charge on the molecule to its frictional coefficient, the sedimentation coefficient is the ratio of buoyant mass to frictional coefficient. Substitution of the above expression for buoyant mass and the Einstein relationship between frictional coefficient and diffusion coefficient [Eq. (4)] into Eq. (24) yields

$$s = MD(1 - \bar{v}\rho)/RT \tag{25}$$

where R is the gas constant. This is the celebrated Svedberg equation.

In a sedimentation-velocity experiment, the ultracentrifuge rotor is operated at speeds up to 70,000 rpm, generating centrifugal fields as high as 500,000 times the force of gravity. Under the action of such an intense force, the macromolecules, which initially were distributed uniformly throughout the solution in the ultracentrifuge cell, are caused to sediment at appreciable rates toward the bottom of the cell (or float to the top if, as in the case of lipoproteins, the macromolecules are less dense than the solvent). As a consequence, a sedimenting boundary is formed between the supernatant in the upper portion of the cell and the underlying "plateau region" in which the concentration of macromolecule is uniform. If the macromolecule is homogeneous, a single such boundary forms and sediments at a rate corresponding to the rate of sedimentation of the macromolecules in the plateau region. The sedimenting boundary is not infinitely sharp, but blurs with time due to translational diffusion. Also, the concentration in the plateau region decreases with time due to radial dilution in the sector-shaped ultracentrifuge cell. These concepts are illustrated diagrammatically in Fig. 8, which presents schematic plots of macromolecular concentration as a function of position in the cell for increasing time of sedimentation, along with the corresponding plots of concentration gradient. The latter are usually recorded photographically in a sedimentation experiment; it is interesting to note that the term "boundary" generally elicits a mathematical image of such concentration or refractive-index gradient curves. The instantaneous distance of the boundary from the center of rotation is called its position. In practice, the position of the boundary $\hat{r}$ is generally taken to be the position of the maximum ordinate of the gradient curve. The rate of sedimentation of the boundary per unit field is the sedimentation coefficient of the macromolecules in the plateau region. Its value may be computed from the slope of a plot of $\ln \hat{r}$ vs. t in accordance with the relation

$$s = \frac{1}{\omega^2 \hat{r}} \frac{d\hat{r}}{dt} = \frac{1}{\omega^2} \frac{d(\ln \hat{r})}{dt} \tag{26}$$

This value of s is dependent upon the experimental conditions of solvent composition, temperature, and macromolecular concentration. With respect

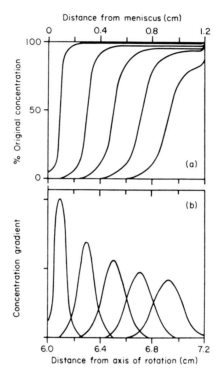

FIG. 8. Schematic plots of (a) macromolecular concentration and (b) concentration gradient as a function of position in the ultracentrifuge cell during the time course of a hypothetical sedimentation-velocity experiment on a single macromolecule. Sedimentation proceeds to right. [From Schachman (35).]

to the latter, it should be noted that the value of s generally increases (often linearly) with decreasing concentration. In order to obtain a value which is a characteristic molecular parameter, it is necessary to perform a set of experiments at progressively decreasing macromolecular concentration in a solvent containing a sufficiently high concentration of supporting electrolyte to depress undesirable charge effects. The several values of s thus obtained are corrected to a reference solvent having the viscosity and density of water at 20°, and then extrapolated to infinite dilution. The extrapolated value of the corrected sedimentation coefficient is a characteristic molecular parameter independent of solute–solute interactions and can be used together with the extrapolated, corrected diffusion coefficient to calculate the anhydrous molecular weight of the macromolecule using the Svedberg equation. In passing, it should be mentioned that information as to the extent of hydration and the shape of the macromolecule also can be obtained by the combined use of sedimentation and diffusion coefficients or of sedimentation coefficient and an independently measured molecular weight.

In addition to the rate of sedimentation of the boundary, its shape and area are also of some interest. Typically, the boundary is a symmetrical bell-shaped curve, although various factors such as strong dependence of sedimentation coefficient on concentration may cause skewing or extreme self-sharpening as, for example, with DNA. Because of these effects, caution must be exercised in using boundary shape or rate of spreading as an index of homogeneity. Factors governing the shape and spreading of the boundary are discussed by Schachman (35) and Dishon and his co-workers (42). Irrespective of shape, the area under the boundary is proportional to the concentration in the plateau region, and thus decreases with time as sedimentation proceeds. The area measured at any given time can be corrected for the radial dilution which has occurred by multiplying by the factor $(\hat{r}/r_m)^2$, where r_m is the distance of the air–liquid meniscus from the center of rotation. The corrected area is proportional to the concentration of macromolecule in the initial solution.

Above, we have considered the sedimentation of a homogeneous macromolecule. If the solution to be analyzed contains not one, but several macromolecules having distinctly different sedimentation coefficients, the resulting sedimentation pattern will show a corresponding number of sedimenting boundaries. Typical patterns are presented in Fig. 9. The resolution of a mixture of two different macromolecules A and B is illustrated diagrammatically in Fig. 10. In this example, B has a larger sedimentation coefficient than A. The slowly sedimenting boundary is formed between pure solvent and solution of pure A, and the faster one between the solution of pure A and a solution containing both A and B. The sedimentation coefficient of each component is computed as described above from the rate of sedimentation of the corresponding boundary. The apparent composition is computed from the areas under the boundaries after correction for dilution in the sector-shaped cell by multiplying each area by its corresponding value of $(\hat{r}/r_m)^2$. The apparent relative proportion of each macromolecule is given by the relative contribution of the corresponding boundary to the total corrected area. However, because of the Johnston–Ogston effect (44), these values are frequently in serious error, the apparent relative proportion of the slower-sedimenting macromolecule being too large and that of the faster one too small. The Johnston–Ogston effect arises from the dependence of the sedimentation coefficient of a given component upon the total macromolecular concentration of the solution through which it is sedimenting. As a result, the concentration of the slower-sedimenting molecule decreases across the faster boun-

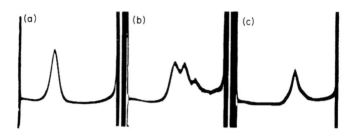

FIG. 9. Sedimentation patterns of different types of purified haptoglobin: (a) phenotype 1–1, $s_{20,w}$ = 4.5 S; (b) phenotype 2–1, $s_{20,w}$ = 4.5, 6, and 7 S; (c) phenotype 2–2, average $s_{20,w}$ = 8 S. Protein concentration, 1.5 g per 100 ml; sedimentation proceeds to right. Compare with starch gel electrophoretic patterns presented in Fig. 6. [From Schultze et al. (43).]

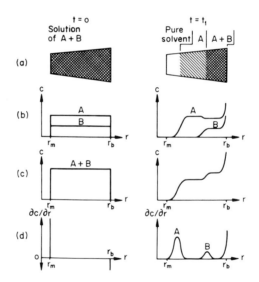

FIG. 10. Resolution of a mixture of two kinds of macromolecules A and B with different sedimentation coefficients by the sedimentation-velocity method: (a) diagrammatic representation of moving boundaries in a sector-shaped analytical cell; (b, c) plots of concentration c against r; (d) plot of concentration gradient $\partial c/\partial r$ against r. Original mixture contains two parts of A and one part of B. Sedimentation coefficient of B is greater than A. Sedimentation proceeds to right. Here r_m and r_b are the distances from the center of rotation of the air–liquid meniscus and the bottom of the cell, respectively. Broadening of the boundaries during sedimentation is due to diffusion. Only very high-molecular-weight substances like DNA and macroglobulins yield almost infinitely sharp boundaries. In general, a solution containing n macromolecules having sufficiently different sedimentation coefficients will give n moving boundaries. [From Cann (43a).]

dary, as illustrated in Fig. 10b. Several procedures are available for obtaining the correct composition; one of which makes use of the fact that the Johnston–Ogston effect is of negligible magnitude at very low concentrations. Analyses are made at progressively decreasing total macromolecular concentration, and the apparent compositions extrapolated to infinite dilution.

We have seen how the sedimentation-velocity method relies on the fact that macromolecules sediment at measurable rates when subjected to the intense centrifugal forces generated at high rotor speeds. In contrast, the sedimentation equilibrium method is used to determine the equilibrium distribution of macromolecules obtained in the low centrifugal fields associated with relatively low rotor speeds, e.g., about 8000 rpm for a protein with a molecular weight of 60,000. During the first stages of a sedimentation-equilibrium experiment, macromolecules are transported in the centrifugal direction due to sedimentation, and the macromolecular concentration decreases at the air–liquid meniscus and increases at the bottom of the ultracentrifuge cell. However, a region devoid of macromolecules is not created as in the sedimentation-velocity method, because of the countertransport of macromolecules in the centripetal direction due to diffusion resulting from the concentration gradient created by the partial sedimentation. As the process proceeds, the plateau region in the center portion of the cell disappears, and there remains only one position in the cell with a concentration equal to the initial concentration. Eventually, an equilibrium state is attained and no further changes in concentration occur with time. At equilibrium, there is no net transport owing to the fact that sedimentation is sufficiently slow as to be exactly counterbalanced by back diffusion. Measurement of the concentration distribution at equilibrium permits computation of the molecular weight. Various procedures have been described for making such calculations. One is to compute a molecular weight using the classical sedimentation-equilibrium equation for ideal solutions,

$$M = \frac{2RT}{(1-\bar{v}\rho)\omega^2} \frac{d(\ln c)}{dr^2} \tag{27}$$

In general, this value for the molecular weight will depart from the correct value due to thermodynamic nonideality of the solution. However, if experiments are made at several concentrations, the correct molecular weight can be obtained by extrapolating a plot of the reciprocal of the computed molecular weights versus concentration to infinite dilution.

Phenomenological Theory of Velocity Sedimentation. As in the case of electrophoresis, sedimentation in the ultracentrifuge is basically a forced-diffusion process in which free diffusion is modified by a centrifugal force which produces a directed flow of the macromolecules. The fundamental differential equation describing this process is variously known as the forced-diffusion, conservation, continuity, or Lamm equation. It is but a mathematical statement of the conservation of mass during sedimentation, and its solution gives the sedimentation pattern. The differential equation can be derived either by application of the thermodynamics of irreversible processes (*36*) or by a kinetic approach (*45*) similar to that used above for electrophoresis. Although the former clearly indicates the conditions for which the equation holds rigorously (i.e., constant partial specific volumes of the components and both s and D referred to a volume fixed frame of reference), we shall make the kinetic derivation, so as to emphasize the basic similarities between the several methods of mass transport. In so doing, one imposes material balance while following the motion of macromolecules during a short time interval as they pass into and out of a thin, cylindrical slice of the ultracentrifuge cell (Fig. 11)

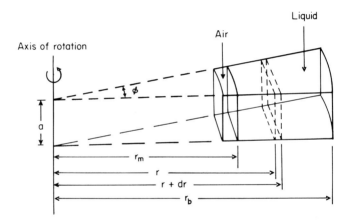

FIG. 11. Schematic diagram showing a thin, cylindrical slice of a sector-shaped ultracentrifuge cell: a, thickness of the cell, i.e., its optical path length; ϕ, angle of the sector; ϕar is the area of the cylindrical surface located at position r; dr is the depth of the cross-sectional slice in the direction of the centrifugal field.

under the influence of the centrifugal force and the random thermal motion of translational diffusion.

The amount of macromolecule entering the slice across its surface at position r in the differential time interval dt due to diffusion is given by Fick's first law as

$$-D\phi ar(\partial c/\partial r)\, dt \tag{28}$$

During the same time interval, the amount of macromolecule

$$-\left(D + \frac{\partial D}{\partial r} dr\right) \phi a(r + dr) \left(\frac{\partial c}{\partial r} + \frac{\partial^2 c}{\partial r^2} dr\right) dt$$

$$= -D\phi a r \frac{\partial c}{\partial r} dt - \left[\frac{\partial}{\partial r}\left(rD \frac{\partial c}{\partial r}\right)\right] \phi a \, dr \, dt \quad (29)$$

leaves the slice by diffusion across the surface at position $r + dr$. The amount of material accumulated in the slice due to diffusion is given by the difference between expressions (28) and (29),

$$\left[\frac{\partial}{\partial r}\left(rD \frac{\partial c}{\partial r}\right)\right] \phi a \, dr \, dt \quad (30)$$

which is a statement of Fick's second law of diffusion in a sector-shaped cell.

The amount of macromolecule simultaneously entering the slice through the surface at r due to sedimentation is given by the product of the concentration, area of the surface, rate of sedimentation of the macromolecules, and the time interval. Since by definition the sedimentation coefficient of a molecule is its rate of sedimentation per unit field, the rate is given by $s\omega^2 r$, and the amount entering the slice becomes

$$c(\phi a r)(s\omega^2 r) \, dt = c\phi a s\omega^2 r^2 \, dt \quad (31)$$

The amount leaving through the surface at $r + dr$ is

$$\left(c + \frac{\partial c}{\partial r} dr\right) \phi a(r + dr) \left(s + \frac{\partial s}{\partial r} dr\right) \omega^2 (r + dr) \, dt$$

$$= c\phi a s\omega^2 r^2 \, dt + \left[\frac{\partial}{\partial r}(cs\omega^2 r^2)\right] \phi a \, dr \, dt \quad (32)$$

The accumulation in the slice due to sedimentation is the difference between expressions (31) and (32),

$$-\left[\frac{\partial}{\partial r}(cs\omega^2 r^2)\right] \phi a \, dr \, dt \quad (33)$$

The total accumulation in the slice due to diffusion and sedimentation, $(\partial c/\partial t)\phi a r \, dr \, dt$, is equal to the sum of expressions (30) and (33),

$$\frac{\partial c}{\partial t} \phi a r \, dr \, dt = \left[\frac{\partial}{\partial r}\left(rD \frac{\partial c}{\partial r}\right)\right] \phi a \, dr \, dt - \left[\frac{\partial}{\partial r}(cs\omega^2 r^2)\right] \phi a \, dr \, dt$$

or

$$\frac{\partial c}{\partial t} = \frac{1}{r} \frac{\partial}{\partial r}\left[\left(D \frac{\partial c}{\partial r} - cs\omega^2 r\right) r\right] \quad (34)$$

which is the desired differential equation for sedimentation.

Since solution of this equation presents formidable mathematical problems, many of the investigations into the sedimentation of reacting systems to be discussed in succeeding chapters are founded on differential equations which make the rectilinear approximation. Not only is rectilinear rather than radial motion of the sedimenting molecules assumed, but the diffusion coefficient and rate of sedimentation v are taken to be constants. Given these approximations, the differential equation for sedimentation becomes

$$\frac{\partial c}{\partial t} = D \frac{\partial^2 c}{\partial x^2} - v \frac{\partial c}{\partial x} \tag{35}$$

which formally is the same as the forced-diffusion equation for ideal electrophoresis [Eq. (19)]. Accordingly, the results of calculations utilizing equations which make the rectilinear approximation are equally applicable to electrophoresis. Conversely, electrophoretic calculations apply to sedimentation to the rectilinear approximation.

Above, we have derived the continuity equation for a homogeneous macromolecule. In the event that the solution contains not one, but a mixture of independent macromolecular species, a continuity equation of the same form can be written for each species:

$$\frac{\partial c_i}{\partial t} = \frac{1}{r} \frac{\partial}{\partial r} \left[\left(D_i \frac{\partial c_i}{\partial r} - c_i s_i \omega^2 r \right) r \right] \tag{36}$$

The several equations can be summed to give

$$\frac{\partial}{\partial t} \sum^i c_i = \frac{1}{r} \frac{\partial}{\partial r} \left[\left(\sum^i D_i \frac{\partial c_i}{\partial r} - \omega^2 r \sum^i c_i s_i \right) r \right] \tag{37}$$

This equation accounting for all the species can, in turn, be written in the form

$$\frac{\partial \tilde{c}}{\partial t} = \frac{1}{r} \frac{\partial}{\partial r} \left[\left(\tilde{D} \frac{\partial \tilde{c}}{\partial r} - \tilde{c} \tilde{s} \omega^2 r \right) r \right] \tag{38}$$

where $\tilde{c} = \Sigma^i c_i$ is the total macromolecular concentration with average transport properties, $\tilde{D} = (\Sigma^i D_i\, \partial c_i/\partial r)/(\Sigma^i \partial c_i/\partial r)$, and $\tilde{s} = (\Sigma^i c_i s_i)/\Sigma^i c_i$. If c_i is expressed in terms of grams per cubic centimeter, $\tilde{s}$ is the weight-average sedimentation coefficient. Such mixtures may give sedimentation patterns showing i sedimenting boundaries. On the other hand, the differences in D_i and s_i may be such that a single composite boundary is observed. Such a composite boundary results from the sum of the gradient curves of each independently sedimenting species in the solution.

Several investigators have addressed themselves from time to time to the solution of the Lamm equation. Faxin (46) first gave an approximate analytical solution which has served as a guide for sedimentation analysis since the early days of the development of the ultracentrifuge. More recently, exact solutions have been obtained, but they are of such complexity as to be virtually useless for computational purposes. Pertinent literature references are given by Dishon and his co-workers (42), who have presented accurate numerical solutions. Here, we are not so concerned with the details of these solutions as with the several relationships which can be obtained from the differential equation without recourse to its solution. These pertain to the definition of boundary position, correction for radial dilution, the moving-boundary equation, and the Johnston–Ogston effect. Each will be considered in turn.

It is almost universal practice to compute the sedimentation coefficient of a macromolecule from the rate of movement of the maximum ordinate of the gradient curve or of the position at which the concentration is one half the value in the plateau region. Justification for these procedures derives (47) from Faxin's approximate solution of the continuity equation; the calculations of Dishon and his co-workers indicate that, for a homogeneous macromolecule, the errors incurred are usually negligible after sufficient time of sedimentation. Goldberg (48) first pointed out, however, that the correct position of a freely sedimenting boundary is actually the square root of its second moment. (A freely sedimenting boundary is one for which the concentration and concentration gradient at the meniscus is zero.) That the rate of movement of the square root of the second moment gives the weight-average sedimentation coefficient in the plateau region can be seen as follows: The second moment $\overline{r^2}$ is given by

$$\overline{r^2} = \int_{r_m}^{r_p} r^2 \frac{\partial \tilde{c}}{\partial r} dr \bigg/ \int_{r_m}^{r_p} \frac{\partial \tilde{c}}{\partial r} dr = r_p^2 - \frac{2}{\tilde{c}_p} \int_{r_m}^{r_p} r\tilde{c} \, dr \qquad (39)$$

where r_p is some arbitrary position in the plateau region and $\tilde{c}_p$ is the plateau concentration. The rate of movement of the second moment is

$$\frac{d\overline{r^2}}{dt} = -\frac{2}{\tilde{c}_p} \int_{r_m}^{r_p} r \frac{\partial \tilde{c}}{\partial t} dr + \frac{2}{\tilde{c}_p^2} \frac{d\tilde{c}_p}{dt} \int_{r_m}^{r_p} r\tilde{c} \, dr \qquad (40)$$

Since $\partial \tilde{c}_p / \partial r = 0$, Eq. (38) tells us that $d\tilde{c}_p/dt = -2\tilde{s}_p \omega^2 \tilde{c}_p$, where $\tilde{s}_p$ is the weight-average sedimentation coefficient in the plateau region. Upon substituting this expression for $d\tilde{c}_p/dt$ and Eq. (38) for $\partial \tilde{c}/\partial t$ into Eq. (40), one obtains

$$\frac{\overline{dr^2}}{dt} = -\frac{2}{\tilde{c}_p}\int_{r_m}^{r_p}\frac{\partial}{\partial r}\left[\left(\tilde{D}\frac{\partial \tilde{c}}{\partial r} - \tilde{c}\tilde{s}\omega^2 r\right)r\right]dr - \frac{4\tilde{s}_p\omega^2}{\tilde{c}_p}\int_{r_m}^{r_p} r\tilde{c}\, dr$$

$$= 2\tilde{s}_p\omega^2\left[r_p^2 - \frac{2}{\tilde{c}_p}\int_{r_m}^{r_p} r\tilde{c}\, dr\right]$$

$$= 2\tilde{s}_p\omega^2 \overline{r^2} \tag{41}$$

which, upon rearrangement, give the desired relation

$$\tilde{s}_p = \frac{1}{\omega^2}\frac{d[\ln(\overline{r^2})^{1/2}]}{dt} \tag{42}$$

If the boundary is not sedimenting freely, its position is given by the square root of

$$\frac{\tilde{c}_m r_m^2}{\tilde{c}_p} + \frac{1}{\tilde{c}_p}\int_{r_m}^{r_p} r^2\frac{\partial \tilde{c}}{\partial r}\, dr \tag{42a}$$

where

$$\tilde{c}_p = \tilde{c}_m + \int_{r_m}^{r_p} (\partial \tilde{c}/\partial r)\, dr \tag{42b}$$

One of the most important applications of ultracentrifugation is to the analysis of biological materials, which are often mixtures of proteins or other macromolecules differing in size, shape, and density. To a first approximation, the area under a given boundary in the sedimentation pattern is proportional to the concentration of the corresponding component in the plateau region. This is rigorously so for a homogeneous macromolecule. However, the plateau concentration decreases with time due to radial dilution in the sector-shaped cell, so that it is necessary to apply a correction to the area under the boundary if one wishes to obtain the concentration c_0 in the original solution. The correction factor is obtained with the aid of the continuity equation as follows: Since $\partial c_p/\partial r = 0$, Eq. (34) tells us that $dc_p/dt = -2\omega^2 s_p c_p$, so that

$$s_p = -\frac{1}{2\omega^2}\frac{d(\ln c_p)}{dt} \tag{43}$$

But s_p is also related to the position of the boundary, by Eq. (26). Equating these two expressions for s_p, we find that $d(\ln \hat{r}) = -\frac{1}{2} d(\ln c_p)$. The latter can be integrated over the time course of sedimentation by recalling that, at the beginning of the experiment, $\hat{r} = r_m$ and $c_p = c_0$:

$$\int_{r_m}^{r_p} d(\ln \hat{r}) = -\frac{1}{2}\int_{c_0}^{c_p} d(\ln c_p) \tag{44}$$

which becomes
$$c_0 = c_p(\hat{r}/r_m)^2 \qquad (45)$$

In other words, the area under the boundary must be multiplied by the square of the ratio of boundary position to meniscus position in order to obtain a direct measure of the concentration in the original solution. Likewise, the area of each boundary shown by a mixture must be multiplied by its corresponding correction factor before computing relative concentrations. The maximum correction amounts to about 35% for a boundary which has sedimented the length of the ultracentrifuge cell. Even after this correction is made, the apparent composition may be found to depend upon total macromolecular concentration due to the Johnston–Ogston effect.

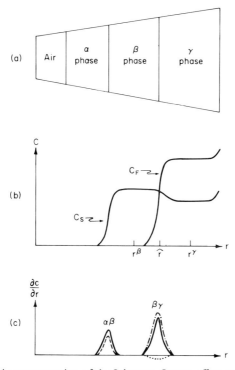

FIG. 12. Schematic representation of the Johnston–Ogston effect operative during sedimentation of a mixture of two components: (a) separation of the liquid contents of the ultracentrifuge cell into three phases as a result of differential sedimentation of the two components; (b) distribution of concentrations of the two components; (c) corresponding gradient curves, where (——) are the observed curves, (– – –) the gradient curve expected for the $\alpha\beta$ boundary on the basis of the concentration of slow component in the γ phase, (– · – · –) the gradient curve expected for the $\beta\gamma$ boundary, (· · · ·) the gradient curve corresponding to the change in concentration of slow component across the $\beta\gamma$ boundary. See text for details.

An understanding of the Johnston–Ogston effect is provided by the moving-boundary equation for sedimentation, which, like the continuity equation from which it derives, is merely a statement of conservation of mass during sedimentation. Consider the sedimentation of a mixture of two macromolecules having different sedimentation coefficients and designated as slow and fast components by a subscript S or F. After sedimentation for a length of time sufficient to cause resolution of the sedimentation pattern into two boundaries, the liquid contents of the ultracentrifuge cell (Fig. 12) may be regarded as having been separated into three phases: (1) an upper, α, phase, composed of pure solvent; (2) a middle, β, phase, containing the slow component at a concentration c_S^β; and (3) a lower, γ, phase, containing both the slow and fast components at the concentrations c_S^γ and c_F^γ, respectively. Separating these phases are the two boundaries: the slower one, designated as the $\alpha\beta$ boundary, sediments at a rate determined by the sedimentation coefficient s_S^β of the slow component in the β phase; and the faster, $\beta\gamma$ boundary, which sediments at a rate determined by the sedimentation coefficient s_F^γ of the fast component in the γ phase. Now, conservation of mass requires that, as the $\beta\gamma$ boundary moves from its position at r^β at time t_1 to r^γ at t_2 (Fig. 12b), the net accumulation of slow component in the volume contained between these two positions must be equal to the integrated product of volume displaced and instantaneous difference in concentration across the boundary. Since the differential accumulation of slow component in the volume element $\phi ar\,dr$ is $(\partial c_S/\partial t)\phi ar\,dr\,dt$, the net amount accumulated as the boundary moves from r^β to r^γ is

$$\int_{t_1}^{t_2}\int_{r^\beta}^{r^\gamma} r(\partial c_S/\partial t)\,dr\,dt \qquad (46)$$

where $\partial c_S/\partial t$ is given by the continuity equation [Eq. (36)] for the slow component. By making this substitution and integrating with respect to position, one obtains

$$\phi a \int_{t_1}^{t_2}\left\{\int_{r^\beta}^{r^\gamma}\left[\frac{\partial}{\partial r}\left(D_S\frac{\partial c_S}{\partial r} - s_S\omega^2 c_S r\right)r\right]dr\right\}dt$$
$$= \phi a \int_{t_1}^{t_2} [s_S^\beta \omega^2 c_S^\beta (r^\beta)^2 - s_S^\gamma \omega^2 c_S^\gamma (r^\gamma)^2]\,dt \qquad (47)$$

The latter is to be equated to the integral

$$\int_{t_1}^{t_2} (\hat{c}_S^\beta - \hat{c}_S^\gamma)\,dV^{\beta\gamma} \qquad (48)$$

where $(\hat{c}_S^\beta - \hat{c}_S^\gamma)$ is the instantaneous change in concentration across the $\beta\gamma$ boundary when located at position $\hat{r}$. The differential volume $dV^{\beta\gamma}$ displaced in the time interval dt is

$$dV^{\beta\gamma} = \phi a\hat{r}\, d\hat{r} = \phi a\omega^2 s^{\beta\gamma}(\hat{r})^2\, dt \tag{49}$$

where the sedimentation coefficient $s^{\beta\gamma}$ of the $\beta\gamma$ boundary is defined by Eq. (26). Accordingly, Eq. (48) becomes

$$\phi a \int_{t_1}^{t_2} \omega^2 s^{\beta\gamma}(\hat{r})^2\, [\hat{c}_S^\beta - \hat{c}_S^\gamma]\, dt \tag{50}$$

Equating integrals (47) and (50), noting that their integrands must also be equal if mass is to be conserved at every instant, and dividing each side by $\hat{r}^2$ yields

$$s_S^\beta c_S^\beta (r^\beta/\hat{r})^2 - s_S^\gamma c_S^\gamma (r^\gamma/\hat{r})^2 = s^{\beta\gamma}[\hat{c}_S^\beta - \hat{c}_S^\gamma] \tag{51}$$

We see that this equation corrects for the radial dilution which occurs during sedimentation of the $\beta\gamma$ boundary, i.e., $c_S^\beta(r^\beta/\hat{r})^2 = \hat{c}_S^\beta$ and $c_S^\gamma(r^\gamma/\hat{r})^2 = \hat{c}_S^\gamma$. If we will remember that the concentrations of the slow component in the β and γ phases are functions of time, we can simplify the notation somewhat and state that, for any given time,

$$s_S^\beta c_S^\beta - s_S^\gamma c_S^\gamma = s^{\beta\gamma}[c_S^\beta - c_S^\gamma] \tag{52}$$

This is the desired moving-boundary equation for sedimentation.

The predictions of the moving-boundary equation become more evident when rearranged to read

$$c_S^\beta / c_S^\gamma = (s^{\beta\gamma} - s_S^\gamma)/(s^{\beta\gamma} - s_S^\beta) \tag{53}$$

It has already been observed that the sedimentation coefficient of a macromolecule decreases with increasing concentration, apparently due to concomitant increase in viscosity, density, and backward flow of the solvent displaced by the sedimenting molecules. Moreover, the sedimentation coefficient is sensitive to the total macromolecular concentration of the solution through which the particular component is sedimenting. Since the total concentration of slow and fast components in the γ phase is greater than the concentration of the slow component in the β phase, $s_S^\gamma < s_S^\beta$. It follows from Eq. (53) that $(c_S^\beta/c_S^\gamma) > 1$, i.e., the concentration of slow component in the β phase is greater than in the γ phase. Furthermore, since the usual correction of c_S^γ for radial dilution gives the concentration of slow component in the initial solution, c_S^β must adjust upward during sedimentation. (It is interesting to note that if this were not the case, slow component would accumulate at the interface between β and γ phases, i.e., within the $\beta\gamma$ boundary itself.) It is evident from

Eq. (53) that the magnitude of the upward adjustment of c_S^β increases as the sedimentation coefficients of the two components approach one another. The described behavior is the Johnston–Ogston effect. Its implications for analysis of mixtures are illustrated in Fig. 12b and c. Because of the upward adjustment of c_S^β, the area of the $\alpha\beta$ boundary is greater than expected from the concentration of the slow component in the plateau region. At the same time, the area under the $\beta\gamma$ boundary is less than expected from the concentration of fast component. This is so because the observed gradient curve is the sum of the gradient of the fast component which disappears across the boundary and the gradient of opposite sense arising from the change in concentration of the slow component across the boundary. The resulting diminution of the $\beta\gamma$ boundary equals exactly the enhancement of the $\alpha\beta$ boundary. Thus, the total area under the two boundaries is correct. (This must be the case, since the total area is determined solely by the total concentration in the γ phase.) Consequently, analysis of the sedimentation pattern in terms of areas overestimates the relative concentration of slow component.

These are not trivial effects, and, in certain instances, can lead to serious error in analysis. Consider, for example, a synthetic mixture containing 20% bovine serum albumin (4.1 S) and 80% fibrinogen (5.4 S) at a total protein concentration of 2.5 g per 100 ml. Sedimentation analysis of the solution (*35*) gave an apparent composition of 50% albumin and 50% fibrinogen. Fortunately, however, such errors become negligible at sufficiently low protein concentration. At low concentrations, the sedimentation coefficient of the slow component in the γ phase is approximately the same as in the β phase, in which case, Eq. (53) tells us that c_S^β is approximately equal to c_S^γ. Accordingly, the correct composition can be obtained by extrapolation of apparent compositions to infinite dilution.

The above considerations illustrate how the continuity equation provides the understanding required for interpretation of sedimentation velocity experiments. Actually, the continuity equation is fundamental to all types of ultracentrifugation. In the case of the Archibald method for determining molecular weight (Appendix I), numerical solution of the continuity equation gives the transient distribution of macromolecule throughout the cell during the approach to equilibrium. At equilibrium, when dc/dt becomes zero, combination of the continuity equation with the thermodynamically deduced sedimentation-equilibrium equation for ideal solutions [Eq. (27)] yields the Svedberg equation [Eq. (25)].

Conditions for Sedimentation Equilibrium of Chemically Reacting Systems. The method of sedimentation equilibrium is enjoying a revival after many years of neglect (*49–51*). Even so, it has been applied to the characterization of relatively few reacting systems of macromolecules. This, despite the fact that the theory is rigorous, involving a minimum of assumptions and approximations readily subject to experimental test. The sedimentation velocity method, on the other hand, is experimentally more rapid and simpler to apply. As pointed out by Steinberg and Schachman (51), the two methods taken together constitute a powerful tool for the analysis of many interacting systems of biological interest.

Earlier in the chapter, sedimentation equilibrium was pictured as the state in which sedimentation is exactly balanced by back-diffusion at all positions in the ultracentrifuge cell. But this kinetic picture is inadequate when considering reacting systems, and appeal must be made to thermodynamics, which rigorously defines the conditions for equilibrium (32, 36, 48, 52, 53).

A system in sedimentation equilibrium may be regarded as composed of a continuous sequence of open phases each of fixed volume[5] and of infinitesimal depth in the direction of the centrifugal field. The temperature, but not the pressure, is uniform throughout the system. Accordingly, at equilibrium the total Helmholtz free energy of the system is a minimum.[6]

Since the volume of the phase bounded by r and $r + dr$ (Fig. 11) is $\phi a r\, dr$, the number of moles of the ith substance (i.e., discrete molecular species) in that phase is $\phi a c_i r\, dr$. Hence, its Gibbs free energy is given by $\phi a \Sigma^l c_i \mu_{ti} r\, dr$, where μ_{ti} is the total chemical potential (Appendix II) and the sum is taken over all l substances comprising the phase. The total Helmholtz free energy of the system A_t is obtained by integrating the Gibbs free energy over all the phases and subtracting therefrom the integrated PV product,

$$A_t = \phi a \int_{r_m}^{r_b} \sum^l c_i \mu_{ti} r\, dr - \phi a \int_{r_m}^{r_b} Pr\, dr = \phi a \int_{r_m}^{r_b} \left(\sum^l c_i \mu_{ti} - P \right) r\, dr \quad (54)$$

At equilibrium A_t is stationary, i.e., $dA_t = 0$ for all permissible variations in the concentrations of the substances.

Consider, first, a system of l substances which sediment independently, in the sense that none are formed out of others. In this event, variations in con-

---

[5] The volume of each phase can be considered as constant when the actual volume change on reaction is a negligible part of the total volume, since the phases are not separated by surfaces of discontinuity (52). In the event that the volume change on reaction is not negligible, a new position variable $\int \rho\, dr$ could be used, in which case, the analogous conditions for equilibrium in the variable would be the same as found herein.

[6] In the most general treatment of this problem, the total internal energy of the system is minimized subject, among other things, to the constraint of constant total entropy. It follows therefrom that uniformity of temperature is a necessary condition for equilibrium. If uniformity of temperature is taken for granted, the additional conditions which are necessary and sufficient for equilibrium may be obtained by minimizing the total Helmholtz free energy.

centration arise solely from free exchange of material between a phase at any position r in the ultracentrifuge cell and adjacent phases. For such a system, variations in the concentrations are subject to the constraint that the total mass of each substance $\mathscr{M}_{ti}$ must be conserved independently. This is expressed mathematically by asserting that each of the integrals

$$\mathscr{M}_{ti} = \phi a \int_{r_m}^{r_b} c_i M_i r \, dr \qquad (i = 1, 2, \ldots, l) \tag{55}$$

must remain constant. That being the case, the integral

$$I = \phi a \int_{r_m}^{r_b} \left(\sum^l c_i \mu_{ti} - \sum^l \lambda_i c_i M_i - P \right) r \, dr \tag{56}$$

where λ_i are Lagrangian undetermined multipliers, is also stationary (54). Moreover, $dI = 0$ for all arbitrary variations of the concentrations independent of mass constraint[7]; and the necessary and sufficient conditions for minimizing the Helmholtz free energy of the system can be ascertained by minimizing the integrand of Eq. (56),

$$\sum_{j=1}^{l} \left\{ \frac{\partial}{\partial c_j} \left[\sum^l c_i \mu_{ti} - \sum^l \lambda_i c_i M_i - P \right] \right\} dc_j = 0 \tag{57}$$

$$\frac{\partial}{\partial c_j} \left[\sum^l c_i \mu_{ti} - \sum^l \lambda_i c_i M_i - P \right] = 0 \qquad (j = 1, 2, \ldots, l) \tag{58}$$

Upon differentiation, Eqs. (58) become

$$\sum_{i=1}^{l} c_i \left(\frac{\partial \mu_{ti}}{\partial c_j} \right)_{T, V, c_i \neq j} + \mu_{tj} - \lambda_j M_j - \left(\frac{\partial P}{\partial c_j} \right)_{T, V, c_i \neq j} = 0 \qquad (j = 1, 2, \ldots, l) \tag{59}$$

where P is the pressure at the position considered. Since the centrifugal contribution to the total potential is independent of concentration, the Gibbs equation tells us that

$$\sum_{i=1}^{l} c_i \left(\frac{\partial \mu_{ti}}{\partial c_j} \right)_{T, V, c_i \neq j} = \left(\frac{\partial P}{\partial c_j} \right)_{T, V, c_i \neq j} \tag{60}$$

at all positions, so that

$$\mu_{tj} = \lambda_j M_j \qquad (j = 1, 2, \ldots, l) \tag{61}$$

---

[7] The method used here is, in principle, the same as that used by Gibbs (52) to reduce the general condition of equilibrium subject to certain constraints to a condition of equilibrium independent of the constraints.

Thus, μ_{tj} is proportional to M_j and the undetermined multipliers are seen to be the total chemical potentials per gram. In other words, at sedimentation equilibrium, the total chemical potential of each substance is constant, i.e., it is the same at every position in the cell.

Now let us consider a system in which only $k = 1, 2, \ldots, l-3$ of the substances sediment independently, the lth substance being a reversible association complex of the $(l-2)$th with the $(l-1)$th substance formed in accordance with the chemical reaction equation

$$\nu_{l-2}\mathscr{G}_{l-2} + \nu_{l-1}\mathscr{G}_{l-1} \rightleftharpoons \mathscr{G}_l \tag{62}$$

In this case also, the total Helmholtz free energy is given by Eq. (54), but the equations of mass constraint must now take into account the chemical reaction. When this is done, variations in the concentrations of all substances can be considered as independent of each other for the purpose of ascertaining the necessary and sufficient conditions for sedimentation equilibrium. Whereas the total mass of each of the k independently sedimenting substances is still given by

$$\mathscr{M}_{tk} = \phi a \int_{r_m}^{r_b} c_k M_k r \, dr \qquad (k = 1, 2, \ldots, l-3) \tag{63}$$

the total masses of the two complexing components are

$$\mathscr{M}_{t,l-2} = \phi a \int_{r_m}^{r_b} [c_{l-2} + \nu_{l-2}c_l] M_{l-2} r \, dr \tag{64}$$

and

$$\mathscr{M}_{t,l-1} = \phi a \int_{r_m}^{r_b} [c_{l-1} + \nu_{l-1}c_l] M_{l-1} r \, dr \tag{65}$$

Since a separate expression for the mass of the complex would be redundant, there is one less equation of mass constraint than in the above case of l independently sedimenting substances. In other words, we are dealing with a system of $l-1$ independent components. The conditions for sedimentation equilibrium are now given by the equations

$$\sum_{j=1}^{l} \left\{ \frac{\partial}{\partial c_j} \left[\sum^{l} c_i \mu_{ti} - \sum^{l-3} \lambda_k c_k M_k - \lambda_{l-2}(c_{l-2} + \nu_{l-2}c_l)M_{l-2} \right. \right.$$
$$\left. \left. - \lambda_{l-1}(c_{l-1} + \nu_{l-1}c_l)M_{l-1} - P \right] \right\} dc_j = 0$$
$$\frac{\partial}{\partial c_j}\left[\sum^{l} c_i\mu_{ti} - \sum^{l-3} \lambda_k c_k M_k - \lambda_{l-2}(c_{l-2} + \nu_{l-2}c_l)M_{l-2}\right.$$
$$\left. - \lambda_{l-1}(c_{l-1} + \nu_{l-1}c_l)M_{l-1} - P \right] = 0 \qquad (j = 1, 2, \ldots, l) \tag{66}$$

These reduce to

$$\sum_{i}^{l} c_i \frac{\partial \mu_{ti}}{\partial c_k} + \mu_{tk} - \lambda_k M_k - \frac{\partial P}{\partial c_k} = 0 \quad (k = 1, 2, \ldots, l-3)$$

$$\sum_{i}^{l} c_i \frac{\partial \mu_{ti}}{\partial c_{l-2}} + \mu_{t,l-2} - \lambda_{l-2} M_{l-2} - \frac{\partial P}{\partial c_{l-2}} = 0,$$

$$\sum_{i}^{l} c_i \frac{\partial \mu_{ti}}{\partial c_{l-1}} + \mu_{t,l-1} - \lambda_{l-1} M_{l-1} - \frac{\partial P}{\partial c_{l-1}} = 0 \quad (67)$$

$$\sum_{i}^{l} c_i \frac{\partial \mu_{ti}}{\partial c_l} + \mu_{tl} - \lambda_{l-2} \nu_{l-2} M_{l-2} - \lambda_{l-1} \nu_{l-1} M_{l-1} - \frac{\partial P}{\partial c_l} = 0$$

Applying the Gibbs equation (60), it follows that

$$\mu_{tk} = \lambda_k M_k \quad (k = 1, 2, \ldots, l-3) \quad (68)$$

$$\mu_{t,l-2} = \lambda_{l-2} M_{l-2} \quad (69)$$

$$\mu_{t,l-1} = \lambda_{l-1} M_{l-1} \quad (70)$$

$$\mu_{tl} = \lambda_{l-2} \nu_{l-2} M_{l-2} + \lambda_{l-1} \nu_{l-1} M_{l-1} \quad (71)$$

Thus, at sedimentation equilibrium the total chemical potential of each molecular species is constant irrespective of whether or not it participates in the chemical reaction. Furthermore, substitution of Eqs. (67) and (70) into Eq. (71) yields

$$\mu_{tl} = \nu_{l-2} \mu_{t,l-2} + \nu_{l-1} \mu_{t,l-1} \quad (72)$$

Inserting the definition of total chemical potential into this expression gives

$$(\mu_l - \nu_{l-2} \mu_{l-2} - \nu_{l-1} \mu_{l-1}) - \tfrac{1}{2} \omega^2 r^2 (M_l - \nu_{l-2} M_{l-2} - \nu_{l-1} M_{l-1}) = 0 \quad (73)$$

and, recalling that $(M_l - \nu_{l-2} M_{l-2} - \nu_{l-1} M_{l-1}) = 0$,

$$\mu_l - \nu_{l-2} \mu_{l-2} - \nu_{l-1} \mu_{l-1} = 0 \quad (74)$$

which will be recognized as the condition for chemical equilibrium. The conclusion reached is that at sedimentation equilibrium the chemical reaction (62) is at equilibrium at every position in the ultracentrifuge cell. Conversely, the condition for chemical equilibrium is the same in the presence as in the absence of the centrifugal force,[8] although the equilibrium concentrations of different molecular species will vary with position due to the centrifugal field.

---

[8] The several integrations could have been carried out for a hypothetical centrifuge cell with the meniscus at the center of rotation.

These conclusions can be generalized to include any reversible reaction, since the relations between the substances participating in the reaction only affect the equations of mass constraint.

Although chemical equilibrium is obtained at every position in the cell, the value of the equilibrium constant need not necessarily be the same at every point. The equilibrium constant will be independent of r only if the molar volume change $\Delta \bar{V}$ of the reaction is zero. This can be seen by considering the reaction $\Sigma^q v_q \mathcal{G}_q = 0$. Since the total chemical potential of each of the substances comprising the system, whether sedimenting independently or in equilibrium with each other, is constant

$$d\mu_q - M_q \omega^2 r \, dr = 0 \tag{75}$$

for all q. If $d\mu_q$ is expressed as an explicit function of position and composition, Eq. (75) can be integrated for ideal solutes (32, 48) to give the familiar logarithmic form of the equation for sedimentation equilibrium

$$\ln\left(\frac{c_\beta}{c_\alpha}\right)_q = \frac{M_q(1 - \bar{v}_q \rho)\omega^2}{2RT}(r_\beta^2 - r_\alpha^2) \tag{76}$$

where c_β and c_α are the concentrations of the qth substance at positions r_β and r_α. The equilibrium constant at r_β is given by

$$K_\beta = (\prod^q c_q^{v_q})_\beta \tag{77}$$

Substituting into this equation the expression for c_q at position β from Eq. (76),

$$K_\beta = (\prod^q c_q^{v_q})_\alpha \exp\left\{-\frac{\omega^2(r_\beta^2 - r_\alpha^2)\rho}{2RT} \sum^q v_q M_q \bar{v}_q\right\}$$

$$= K_\alpha \exp\left\{-\frac{\omega^2(r_\beta^2 - r_\alpha^2)\rho}{2RT} \Delta \bar{V}\right\} \tag{78}$$

where K_α is the equilibrium constant at r_α. If $\Delta \bar{V} = 0$, $K_\beta = K_\alpha$.

Taking note that the factor $\omega^2(r_\beta^2 - r_\alpha^2)\rho/2$ in Eq. (78) is the difference in pressure (32) between positions r_β and r_α, it becomes apparent that Eq. (78) is simply the integrated form of the reaction isotherm

$$\left(\frac{\partial \ln K}{\partial P}\right)_T = -\frac{\Delta \bar{V}}{RT} \tag{79}$$

Finally, it is instructive to note that the equilibrium distribution in the centrifuge cell of each of the various substances participating in the reversible association–dissociation reaction $\Sigma^q v_q \mathcal{G}_q = 0$ is identical to that obtained by the corresponding substance in a synthetic, nonreacting system of composition identical to that of the reacting system in the absence of the centrifugal field. The synthetic system is constructed from q hypothetical, nonreacting substances each having the same molecular weight, partial specific volume, and activity coefficient as the corresponding reactant or product, $\mathscr{G}_q$. These are mixed such that the total concentration of the solution and the concentration of each substance are the same as in the reacting system at equilibrium in the absence of the field. At sedimentation equilibrium, these two systems are indistinguishable, since the conditions for sedimentation equilibrium require equilibrium with respect to all factors—thermal, mechanical, and chemical. The two systems can be distinguished, however, by performing sedimentation-equilibrium experiments on their serial dilutions. As the total concentration of the reacting system is lowered, its weight-average molecular weight will decrease due to mass action. Depending on the concentration range employed, this decrease should be readily distinguishable from the drift in the apparent molecular weight of the synthetic system due to the usual thermodynamic nonideality.

APPENDIX I: SOME OTHER METHODS OF SEDIMENTATION ANALYSIS

Band Sedimentation. In band sedimentation (*40*), a thin lamella of a dilute solution of nucleic acid, protein, or virus is layered on to a denser liquid (solvent containing more concentrated electrolyte or D_2O) under the influence of the centrifugal field in a special sector-shaped cell. Macromolecules of the same sedimentation coefficient then sediment through the liquid in a narrow band. Sedimentation of a mixture of noninteracting macromolecules having different sedimentation coefficients results in a corresponding number of bands. The density gradients necessary to stabilize the system against convection are generated during the experiment by diffusion of small molecules between the lamella and the bulk solution, and, in some cases, by sedimentation of the small molecules in the bulk solution. The bands are recorded using ultraviolet absorption optics. In contrast to analytical velocity sedimentation, which is a moving-boundary method, band sedimentation permits complete physical separation of different kinds of macromolecules, thereby permitting analysis without the complication of the Johnston–Ogston effect. Sedimentation coefficients may be evaluated from the motion of the bands.

Archibald Approach to Sedimentation-Equilibrium Method. Because of the long times required to attain sedimentation equilibrium of macromolecules of the size of many proteins, considerable effort has been directed toward theoretical analysis of the intermediate stages while the solute is being redistributed in the ultracentrifuge cell. Such considerations led Archibald (*55*) to the realization that molecular weights can be determined from a knowledge of the instantaneous concentration distribution in the regions near the meniscus and the bottom of the centrifuge cell during the approach to equilibrium in experiments of short duration. So great has been the success and breadth of the Archibald method, that it has become one of the most widely used methods for the determination of molecular weights of both large and small molecules. The theoretical basis of the method resides in the fact that, at all times during an experiment, there is no net flow of solute through the surfaces at the meniscus or the bottom of the cell. The amount of solute crossing a surface at position r in the cell due to diffusion and sedimentation during the differential time interval dt is found, by adding Eqs. (28) and (31), to be

$$\phi ar[cs\omega^2 r - D\, \partial c/\partial r]\, dt$$

Now, the factor in brackets must vanish at the meniscus and bottom of the cell, so that

$$c_m s\omega^2 r_m = D(\partial c/\partial r)_m \quad \text{and} \quad c_b s\omega^2 r_b = D(\partial c/\partial r)_b$$

Combining these relationships with the Svedberg equation (25) gives the following expressions for the molecular weight at the meniscus and the bottom:

$$M_m = \frac{RT}{(1 - \bar{v}\rho)\omega^2} \frac{(\partial c/\partial r)_m}{r_m c_m}$$

and

$$M_b = \frac{RT}{(1 - \bar{v}\rho)\omega^2} \frac{(\partial c/\partial r)_b}{r_b c_b}$$

If the substance under investigation is homogeneous, $M_m = M_b$. Thus, not only is the method rapid and accurate, but, more importantly, it provides a sensitive method for detecting heterogeneity. Experimental conditions are selected such that the concentration ratio c_b/c_m, had equilibrium been attained, would have values ranging from 2:1 to 5:1. Values of $\partial c/\partial r$, c_m, and c_b are determined from Schlieren patterns using computational methods described by Schachman (*35*).

Density-Gradient Equilibrium Sedimentation. As described in the text, the Svedberg equation provides the theoretical bases for differential centrifugation of substances having different molecular weights and/or shapes in a solvent less dense than the solute molecules (velocity sedimentation). It also shows that differences in solute density $1/\bar{v}$ can be utilized to achieve separation; e.g., consider a mixture of protein and lipoprotein in a salt solution whose density is less than that of the protein, but greater than the lipoprotein; under the action of a centrifugal field, the protein will sediment to the bottom of the centrifuge column while the lipoprotein floats to the top. If the density of the solvent is equal to that of the solute, the solute molecules neither sediment nor float. These principles are utilized in the separation of deoxyribonucleic acids (DNA's) by the method of sedimentation equilibrium in a field-generated density gradient of cesium chloride (*39*). Centrifugation of a concentrated cesium chloride solution results in an appreciable concentration gradient of cesium chloride, and thus a density gradient. When DNA is present, the macromolecules redistribute themselves in the density gradient. The DNA molecules sediment, float upward, or remain stationary, depending upon the local density. At equilibrium, DNA's of different densities are narrowly banded at different positions in the ultracentrifuge column. Each DNA bands at a position where the density of the cesium chloride solution and effective density of the macromolecule in that solution are equal. The thermodynamic conditions for equilibrium are the same as for classical sedimentation equilibrium. This method was originally devised to elucidate the mechanism of DNA replication, and is finding important applications in nucleic acid chemistry. It has also been applied to the analysis of lipoproteins and proteins in sucrose, cesium chloride, and rubidium chloride gradients.

Zone-Velocity Sedimentation in a Preformed Density Gradient. In zone-velocity sedimentation through a preformed density gradient (*37, 38*), a small volume of nucleic acid or protein solution is layered on top of a fairly small, continuous density gradient. The usually linear gradient is preformed from light and heavy solutions with some mechanical device using gradient-forming materials such as sucrose or cesium chloride. Experiments are performed in cylindrical Lusteroid tubes using a swinging bucket rotor in a preparative ultracentrifuge such as the Spinco Model L. Under the action of the centrifugal field, macromolecules with the same sedimentation coefficient sediment as a more or less narrow zone through the gradient, which stabilizes the systems against convection. A mixture of several different kinds of molecules yields a corresponding number of discrete zones. Rotor speeds and times of sedimentation are chosen to cause separation of the zones before the most rapidly sedimenting one reaches the bottom of the tube. Upon cessation of centrifugation the contents of the tube are fractionated by any one of several methods, e.g., piercing the bottom of the tube and collecting drops. The fractions are then analyzed for radioactivity, biological activity, chemical properties, optical density, etc., and sedimentation patterns constructed by plotting the results of the analyses against distance from the meniscus or center of rotation, volume, number of drops, or fraction number. Thus, zone-velocity sedimentation provides both analysis and fractionation of a mixture in a single operation. Furthermore, a particular biologically active substance in a crude extract may be localized by its activity and characterized at concentrations so low as to be undetectable by the most sensitive analytical sedimentation procedures. Sedimentation coefficients can be computed directly from the rates of migration of the zones and knowledge of the viscosity and density at each point in the gradient. Alternately, a standard well-characterized protein can be added to the original protein mixture. The sedimentation coefficient of the unknown is then given by the product of that of the standard and the ratio of the distances sedimented by the unknown and the standard (*38*).

APPENDIX II: TOTAL CHEMICAL POTENTIAL

μ_{ti} is called the total chemical potential of substance i and is defined as $\mu_{ti} = \mu_i - \frac{1}{2}M_i\omega^2 r^2$, where μ_i is the ordinary chemical potential and $-\frac{1}{2}M_i\omega^2 r^2$ is the centrifugal potential. The ordinary chemical potential is defined as

$$\mu_i = \left(\frac{\partial E}{\partial n_i}\right)_{S,V,n_{j\neq i}} = \left(\frac{\partial G}{\partial n_i}\right)_{T,P,n_{j\neq i}} = \left(\frac{\partial A}{\partial n_i}\right)_{T,V,n_{j\neq i}} = \left(\frac{\partial H}{\partial n_i}\right)_{S,P,n_{j\neq i}}$$

and is a function of temperature T, pressure P, and composition. Here, E is the internal energy of an open, restricted system; G its Gibbs free energy; A the Helmholtz free energy; H the enthalpy; S the entropy; V is volume; and n_i is mole number. An open thermodynamic system is one which may exchange mass as well as work and heat with its environment; a restricted system is one which is not subject to the action of an external force, e.g., centrifugal force. For an infinitesimal reversible change in an open, restricted system, the change in Helmholtz free energy, for example, is

$$dA = -S\,dT - P\,dV + \sum^i \mu_i\,dn_i$$

In an open system unrestricted in that it is subject to the action of a centrifugal force, account must be taken of the additional change in energy associated with the transfer of material in the centrifugal field. When this is done, the centrifugal potential adds to the ordinary chemical potential in a very natural way. Thus, for an infinitesimal reversible change in such a system, the change in Helmholtz free energy is

$$dA = -S\,dT - P\,dV + \sum^i \mu_i\,dn_i + \Phi\,d\mathcal{M}$$

where $\Phi\,d\mathcal{M}$ is the work required to transfer the mass $d\mathcal{M} = \Sigma^i M_i\,dn_i$ through a change in centrifugal potential $\Phi = -\int_0^r \omega^2 r\,dr = -\frac{1}{2}\omega^2 r^2$. Accordingly,

$$dA = -S\,dT - P\,dV + \sum^i (\mu_i - \tfrac{1}{2}M_i\omega^2 r^2)\,dn_i$$

Since

$$dA = \left(\frac{\partial A}{\partial T}\right)_{V,n_i} dT + \left(\frac{\partial A}{\partial V}\right)_{T,n_i} dV + \sum^i \left(\frac{\partial A}{\partial n_i}\right)_{T,V,n_{j\neq i}} dn_i$$

it follows that

$$\left(\frac{\partial A}{\partial n_i}\right)_{T, V, n_{j \neq i}} = \mu_i - \tfrac{1}{2} M_i \omega^2 r^2 = \mu_{ti}.$$

which also clarifies the meaning of the term "total chemical potential." Likewise,

$$\left(\frac{\partial E}{\partial n_i}\right)_{S, V, n_{j \neq i}} = \left(\frac{\partial G}{\partial n_i}\right)_{T, P, n_{j \neq i}} = \left(\frac{\partial H}{\partial n_i}\right)_{S, P, n_{j \neq i}} = \mu_{ti}$$

for such a system. Above, we have made a distinction between the ordinary chemical potential and the total potential in a system subject to the action of a centrifugal force. In general, other forces (of which centrifugal force is a particular example) may act on the system, in which case, the total chemical potential is the sum of the ordinary chemical potential and the appropriate external potentials. Finally, the Helmholtz free energy of an open unrestricted system is $-PV + \Sigma^i \mu_{ti} n_i$; and for a change in state at constant temperature and volume, the Gibbs equation defines the change in pressure, $\Sigma^i n_i d\mu_i = V dP$. The most general form of the Gibbs equation is $\Sigma^i n_i d\mu_i = -S dT + V dP$. At constant temperature and pressure, the Gibbs equation reduces to the Gibbs–Duhem equation, $\Sigma^i n_i d\mu_i = 0$.

REFERENCES

1. A. Tiselius, *Z. Physik. Chem.* (*Leipzig*) **124**, 449 (1926).
2. L. G. Longsworth and D. A. MacInnes, *J. Gen. Physiol.* **25**, 507 (1942).
3. F. F. Reuss, *Mem. Imp. Naturalists Moscou* **2**, 327 (1809)
4. O. Lodge, *Rept. Brit. Assoc.* **56**, 369 (1886).
5. D. A. MacInnes and L. G. Longsworth, *Chem. Rev.* **11**, 171 (1932).
6. H. Picton and S. E. Linder, *J. Chem. Soc.* **61**, 148 (1942).
7. A. Tiselius, *Trans. Faraday Soc.* **33**, 524 (1937).
8. L. G. Longsworth, *Ind. Eng. Chem. Anal. Ed.* **18**, 219 (1946).
9. D. Cremer and A. Tiselius, *Biochem. Z.* **320**, 273 (1950).
10. E. L. Durrum, *J. Am. Chem. Soc.* **72**, 2943 (1950).
11. K. Kraus and G. Smith, *J. Am. Chem. Soc.* **72**, 4329 (1950).
12. H. McDonald, M. Urbin and M. Williamson, *Science* **112**, 227 (1950).
13. F. Turba and H. Emenkal, *Naturwissenschaften* **37**, 93 (1950).
14. W. Grassman and K. Hannig, *Naturwissenschaften* **37**, 93 (1950).
15. H. A. Abramson, L. S. Moyer and M. H. Gorin, "Electrophoresis of Proteins and the Chemistry of Cell Surfaces." Reinhold, New York, 1942.
16. H. Muller, *in* "Proteins, Amino Acids and Peptides as Ions and Dipolar Ions" (E. J. Cohn and J. T. Edsall, eds.), Chapter 25. Reinhold, New York, 1943.
17. J. Th. G. Overbeek, *Advan. Colloid Sci.* **3**, 97 (1950).

References

18. J. Th. G. Overbeek and P. H. Wiersema, in "Electrophoresis Theory, Methods, and Applications" (M. Bier, ed.), Vol. II, Chapter 1. Academic Press, New York, 1967.
18a. D. C. Henry, Proc. Roy. Soc. (London) **A133**, 106 (1931).
19. P. D. Ross, Biopolymers **2**, 9 (1964).
20. B. M. Olivera, P. Baine and N. Davidson, Biopolymers **2**, 245 (1964).
21. A. E. Alexander and P. Johnson, in "Colloid Science," Vol. 1, pp. 180–186. Oxford Univ. Press, London and New York, 1949.
22. L. G. Longsworth, in "Electrophoresis Theory, Methods and Applications" (M. Bier, ed.), Chapters 3 and 4. Academic Press, New York, 1959.
23. J. R. Cann, in "Treatise on Analytical Chemistry" (I. M. Kolthoff and P. J. Elving, ed.), Vol. 2, Part 1, Chapter 28. Wiley (Interscience), New York, 1961.
23a. J. R. Cann, J. Phys. Chem. **63**, 210 (1959).
23b J. R. Cann, J. Chem. Phys. **33**, 1410 (1960).
23c. V. P. Dole, J. Am. Chem. Soc. **67**, 1119 (1945).
24. H. G. Kunkel and A. Tiselius, J. Gen. Physiol. **35**, 89 (1951).
25. H. Waldmann-Meyer, J. Biol. Chem. **235**, 3337 (1960).
26. H. Waldmann-Meyer, Chromatog. Rev. **5**, 1 (1963).
27. H. Waldmann-Meyer, Methods Biochem. Analy. **13**, 47 (1965).
28. H. Waldmann-Meyer and K. Schilling, Acta Chem. Scand. **13**, 13 (1959).
29. J. R. Cann and M. H. Loveless, J. Allergy **28**, 379 (1957).
30. R. J. Block, E. L. Durrum and G. Zweig, "A Manual of Paper Chromatography and Paper Electrophoresis," 2nd ed. Academic Press, New York, 1958.
31. H. Harris, Brit. Med. Bull. **17**, 217 (1961).
32. T. Svedberg and K. O. Pedersen, "The Ultracentrifuge." Oxford Univ. Press, London and New York, 1940; Johnson Reprint Corp., New York.
33. F. Sanger, Science **129**, 1340 (1959).
34. J. C. Kendrew, Sci. Am. **205**, 96 (1961).
35. H. K. Schachman, "Ultracentrifugation in Biochemistry." Academic Press, New York, 1959.
36. H. Fujita, "Mathematical Theory of Sedimentation Analysis." Academic Press, New York, 1962.
37. C. deDuve, J. Berthet and H. Beaufay, Prog. Biophys. Biophys. Chem. **9**, 326 (1959).
38. R. G. Martin and B. N. Ames, J. Biol. Chem. **236**, 1372 (1961).
39. M. Meselson, F. W. Stahl and J. Vinograd, Proc. Natl. Acad. Sci. U.S. **43**, 581 (1957); **44**, 671 (1958).
40. J. Vinograd, R. Brunner, R. Kent and J. Weigle, Proc. Natl. Acad. Sci. U.S. **49**, 902 (1963).
41. D. A. Yphantis, Biochemistry **3**, 297 (1964).
42. M. Dishon, G. H. Weiss and D. A. Yphantis, Biopolymers **5**, 697 (1967).
43. H. E. Schultze, H. Haupt, K. Heide and N. Heimburger, Clin. Chim. Acta **8**, 207 (1963).
43a. J. R. Cann, in "The Encyclopedia of Biochemistry" (R. J. Williams and E. M. Lansford, Jr., eds.), p. 210. Reinhold, New York, 1967.
44. J. P. Johnston and A. G. Ogston, Trans. Faraday Soc. **42**, 789 (1946).
45. O. Lamm, Z. Physik. Chem. (Leipzig) **A143**, 177 (1929); Arkiv. Mat. Astron. Fysik **21B**, No. 2 (1929).
46. H. Faxen, Arkiv Mat. Astron. Fysik **21B**, No. 3 (1929).
47. O. Lamm, Z. Physik. Chem. (Leipzig) **A143**, 177 (1929).
48. R. J. Goldberg, J. Phys. Chem. **57**, 194 (1953).
49. E. G. Richards, D. C. Teller and H. K. Schachman, Biochemistry **7**, 1054 (1968).
50. D. Roark and D. A. Yphantis, Ann. N.Y. Acad. Sci. **164**, 245, (1969).

51. I. Z. Steinberg and H. K. Schachman, *Biochemistry* **5,** 3728 (1966).
52. J. W. Gibbs, "The Scientific Papers of J. Willard Gibbs," Vol. 1, pp. 62–79, 144–150. Yale Univ. Press, New Haven, Connecticut, 1948.
53. J. W. Williams, K. E. Van Holde, R. L. Baldwin and H. Fujita, *Chem. Rev.* **58,** 715 (1958).
54. H. Margenau and G. M. Murphy, "The Mathematics of Physics and Chemistry," p. 204. Van Nostrand, Princeton, New Jersey, 1943.
55. W. J. Archibald, *J. Phys. Colloid Chem.* **51,** 1204 (1947).

II / WEAK-ELECTROLYTE MOVING-BOUNDARY THEORY

In our previous discussion of the principles of electrophoresis, the conservation equation [Eq. (18)] was derived for a solution containing a single macromolecular ion. Actually, proteins are weak, amphoteric electrolytes, and a solution of an inherently homogeneous protein contains a multiplicity of macromolecular ionic species PH_n participating in acid–base reactions of the type $PH_n \rightleftharpoons PH_{n-1} + H^+$. Each species possesses a characteristic net charge, and thus electrophoretic mobility, and also a characteristic diffusion coefficient. Description of the electrophoretic behavior of such a system must take cognizance of the adjustment of acid–base equilibria which occurs within the moving boundary during differential transport of the various species. When this is done, the electrophoretic patterns are obtained as mathematical solutions of a set of partial differential equations which is an expression of the overall conservation of mass during transport caused by diffusion and the external electrical force acting on the macromolecular ions and the hydrogen ion. Each equation of the set describes the mass transport and chemical interconversion for one of these reactants.

Mathematical formulation of these ideas leads in a natural way to a statement of the weak-electrolyte moving-boundary theory developed by Svensson (1) and by Alberty and his co-workers (2–5). This theory provides the basis for quantitative analysis of the electrophoretic and sedimentation patterns shown by certain types of biologically interesting, reversibly reacting systems, e.g., macromolecule–small-ion and enzyme–macromolecular substrate interactions. Before proceeding, however, it is helpful to illustrate the derivation and mathematical manipulation of the appropriate transport-interconversion equations by considering what is perhaps the simplest of macromolecular interactions—namely, isomerization.

MACROMOLECULAR ISOMERIZATION

Let us suppose that a given protein in solution can exist in two interconvertible states $A \underset{k_2}{\overset{k_1}{\rightleftharpoons}} B$, in which it migrates with electrophoretic mobilities μ_1 and μ_2 and diffuses with diffusion coefficients D_1 and D_2. The specific rates of interconversion are k_1 and k_2; the concentrations of A and B are designated by c_1 and c_2. The transport-interconversion equation for each isomer is derived in the same way as Eq. (18) except that account is taken not only of diffusion and electromigration into and out of the thin cross-sectional slice of the electrophoresis column, but also of the appearance or disappearance of the particular isomer due to the isomerization reaction.

Consider, for example, the isomer A. Its accumulation in the slice due to diffusion is $(\partial/\partial x)(D_1 \, \partial c_1/\partial x)\mathscr{A}\,dx\,dt$; due to electromigration it is $-[\partial(\mu_1 E c_1)/\partial x]\mathscr{A}\,dx\,dt$; and due to the reversible isomerization reaction it is $R_1 \mathscr{A}\,dx\,dt$, where $R_1 = -k_1 c_1 + k_2 c_2$ is the net rate of formation of A from B. The total accumulation of A, $(\partial c_1/\partial t)\mathscr{A}\,dx\,dt$, is equal to the sum of the diffusion, electromigration, and reaction terms

$$\frac{\partial c_1}{\partial t}\mathscr{A}\,dx\,dt = \frac{\partial}{\partial x}\left(D_1 \frac{\partial c_1}{\partial x}\right)\mathscr{A}\,dx\,dt - \frac{\partial(\mu_1 E c_1)}{\partial x}\mathscr{A}\,dx\,dt + R_1 \mathscr{A}\,dx\,dt$$

or

$$\frac{\partial c_1}{\partial t} = \frac{\partial}{\partial x}\left(D_1 \frac{\partial c_1}{\partial x}\right) - \frac{\partial(\mu_1 E c_1)}{\partial x} - k_1 c_1 + k_2 c_2 \qquad (80)$$

which is the desired transport-interconversion equation for this isomer. Likewise, for isomer B,

$$\frac{\partial c_2}{\partial t} = \frac{\partial}{\partial x}\left(D_2 \frac{\partial c_2}{\partial x}\right) - \frac{\partial(\mu_2 E c_2)}{\partial x} + k_1 c_1 - k_2 c_2 \qquad (81)$$

If it is assumed that electrophoresis is ideal and that the diffusion coefficients and electrophoretic mobilities are independent of concentration, Eqs. (80) and (81) simplify to

$$\frac{\partial c_1}{\partial t} = D_1 \frac{\partial^2 c_1}{\partial x^2} - \mu_1 E \frac{\partial c_1}{\partial x} - k_1 c_1 + k_2 c_2 \qquad (82\text{a})$$

$$\frac{\partial c_2}{\partial t} = D_2 \frac{\partial^2 c_2}{\partial x^2} - \mu_2 E \frac{\partial c_2}{\partial x} + k_1 c_1 - k_2 c_2 \qquad (82\text{b})$$

These two differential equations constitute the set whose solution gives the electrophoretic patterns of an isomerizing system. It is evident from inspection that the nature of the patterns will be sensitive to the rates of interconversion of the isomers. Three cases of interest may be discerned. The first is the limiting case in which the rates of interconversion are sufficiently slow that significant reequilibration does not occur during the course of the electrophoretic experiment. In that event, one can neglect the rate terms in Eqs. (82a) and (82b), which then reduce to two uncoupled equations each analogous to the conservation equation [Eq. (18)] for a single macromolecular ion. In other words, the system will behave like a mixture of two noninteracting macromolecules. The patterns will resolve into two moving boundaries corresponding to the isomers A and B; and conventional analysis of the patterns yields their mobilities and concentrations in the initial equilibrium mixture. The situation is considerably more complex, however, when the rates of interconversion are comparable to the rates of diffusion and electrophoretic separation of the isomers. Under these circumstances, the patterns will show three peaks. This case will be dealt with in subsequent chapters. Here, we are concerned primarily with a second limiting case in which the rates of interconversion are so very much larger than the rates of diffusion and electrophoresis that local equilibrium at every instant can be assumed. We proceed as follows:

Addition of Eqs. (82a) and (82b) gives

$$\frac{\partial(c_1 + c_2)}{\partial t} = D_1 \frac{\partial^2 c_1}{\partial x^2} - \mu_1 E \frac{\partial c_1}{\partial x} + D_2 \frac{\partial^2 c_2}{\partial x^2} - \mu_2 E \frac{\partial c_2}{\partial x} \tag{83}$$

which is a statement of conservation of total protein under the action of diffusion and electrophoretic transport. Given the assumption of instantaneous reequilibration during differential transport of the isomers,

$$c_2 = K c_1 \tag{84}$$

where K is the equilibrium constant for isomerization. Since c_2 is just equal to c_1 multiplied by a constant, Eq. (83) can be written in a form in which the variable $\bar{c} = c_1 + c_2$ appears with concentration averaged diffusion coefficient and mobility, namely,

$$\partial \bar{c}/\partial t = \bar{D} \, \partial^2 \bar{c}/\partial x^2 - \bar{\mu} E \, \partial \bar{c}/\partial x$$

$$\bar{D} = \alpha_1 D_1 + \alpha_2 D_2 \quad \text{and} \quad \bar{\mu} = \alpha_1 \mu_1 + \alpha_2 \mu_2 \tag{85}$$

$$\alpha_1 = 1/(1 + K) \quad \text{and} \quad \alpha_2 = K/(1 + K)$$

The form of this equation is the same as the forced-diffusion equation [Eq. (19)] for a single, noninteracting macromolecular ion. Accordingly, in this limit, the two isomers behave as a single macromolecular ion with average transport properties, and the electrophoretic patterns show a single Gaussian-shaped boundary.

The quantity $\bar{c}$ is called the constituent concentration of the isomerizing protein. In general, the constituent concentration of a component of a system is the total concentration of that component in all its forms; e.g., in the treatment of sedimentation equilibrium of reacting systems presented in Chapter I, the constituent concentration of the $(l-2)$th component in a given phase appears in Eq. (64) as the factor $c_{l-2} + v_{l-2}c_l$. The concentration-averaged diffusion coefficient and mobility, $\bar{D}$ and $\bar{\mu}$, are called constituent diffusion coefficient and constituent mobility. Tiselius (5a) first pointed out that a component consisting of several forms with different mobilities in equilibrium with each other will generally migrate as a uniform component with a mobility $\bar{\mu} = \Sigma_{i=0}^{n} \alpha_i \mu_i$, where α_i is the fraction of the component with mobility μ_i, provided the time of existence of each form is small in comparison with the duration of the experiment. These are powerful concepts for the analysis of certain types of interacting systems, and are fundamental to the weak-electrolyte moving-boundary theory which is elaborated below.

WEAK-ELECTROLYTE MOVING BOUNDARY

Returning to our original considerations of the actual state of an inherently homogeneous protein (designated as the constituent $\mathscr{P}$) in solution, let us represent the protein as an ampholyte containing $2v$ acid groups; e.g., v electrically neutral acid groups (e.g., carboxyl groups of aspartic acid residues and phenolic hydroxy groups of tyrosine), and v cationic acid groups (e.g., ammonium groups of lysine and protonated imadazole groups of histidine). This molecule is capable of forming a series of $2v + 1$ macromolecular ions $PH_{2v}, \ldots, PH_n, \ldots, P$ with respective electrical charges $n - v$ and mobilities μ_n. (There are, of course, a total of 2^{2v} distinct macromolecular ions; but, with respect to charge, these constitute only $2v + 1$ classes.) The dissociation equilibria are described by the following set of $2v$ equations and their respective acid dissociation constants:

$$\begin{aligned}
PH_{2v} &\rightleftharpoons PH_{2v-1} + H^+ \; ; & K_1 \\
&\vdots \\
PH_n &\rightleftharpoons PH_{n-1} + H^+ \; ; & K_{2v-n+1} \\
&\vdots \\
PH &\rightleftharpoons P + H^+ \; ; & K_{2v}
\end{aligned} \tag{86}$$

The concentration c_n of the ion PH_n is given by (6)

$$c_n = \bar{c}_{\mathscr{P}} K_n a_{H^+}^n / G(a_{H^+}) = \alpha_n \bar{c}_{\mathscr{P}}$$

$$G(a_{H^+}) = \sum_{k=0}^{2v} K_n a_{H^+}^k, \qquad K_n = \gamma_n^{-1} \prod_{s=1}^{n} K_{2v-s+1}^{-1}, \qquad \bar{c}_{\mathscr{P}} = \sum_{n=0}^{2v} c_n \tag{87}$$

where a_{H^+} is the hydrogen ion activity; γ_n is the activity coefficient of PH_n; α_n is the fraction of the ampholyte existing in the form PH_n; and $\bar{c}_{\mathscr{P}}$ is the constituent concentration of the ampholyte. The net charge $\bar{z}$ of the ampholyte is given by

$$\bar{z} = \sum_{n=0}^{2v} \alpha_n(n - v) \tag{88}$$

and its constituent mobility by

$$\bar{\mu}_{\mathscr{P}} = \sum_{n=0}^{2v} \alpha_n \mu_n \tag{89}$$

Thus, we see that even the simplest protein in solution is an interacting system. Accordingly, its electrophoretic behavior is described by a set of transport-interconversion equations analogous to those derived above for isomerization, namely,

$$\frac{\partial c_n}{\partial t} = \frac{\partial}{\partial x}\left(D_n \frac{\partial c_n}{\partial x}\right) - \frac{\partial(\mu_n E c_n)}{\partial x} + R_n \quad (n = 0, 1, \ldots, 2v) \tag{90}$$

$$\frac{\partial c_{H^+}}{\partial t} = \frac{\partial}{\partial x}\left(D_{H^+} \frac{\partial c_{H^+}}{\partial x}\right) - \frac{\partial(\mu_{H^+} E c_{H^+})}{\partial x} + R_{H^+} \tag{91}$$

By summing Eq. (90) and taking cognizance of the fact that conservation of mass requires $\Sigma_{n=0}^{2v} R_n = 0$ and that acid–base reactions are extremely rapid, we obtain

$$\frac{\partial \bar{c}_{\mathscr{P}}}{\partial t} = \frac{\partial}{\partial x}\left(\bar{D}_{\mathscr{P}} \frac{\partial \bar{c}_{\mathscr{P}}}{\partial x}\right) - \frac{\partial(\bar{\mu}_{\mathscr{P}} E \bar{c}_{\mathscr{P}})}{\partial x}$$

$$\bar{D}_{\mathscr{P}} = \sum_{n=0}^{2v}\left(D_n \frac{\partial c_n}{\partial x}\right) \bigg/ \sum_{n=0}^{2v} \frac{\partial c_n}{\partial x} \tag{92}$$

which is a statement of conservation of total protein during diffusion and electromigration. Similarly, for the hydrogen ion,

$$\frac{\partial \bar{c}_{H^+}}{\partial t} = \frac{\partial}{\partial x}\left(\bar{D}_{H^+} \frac{\partial \bar{c}_{H^+}}{\partial x}\right) - \frac{\partial(\bar{\mu}_{H^+} E \bar{c}_{H^+})}{\partial x} \tag{93}$$

in which appear the constituent quantities

$$\bar{c}_{H^+} = c_{H^+} + \sum_{n=0}^{2v} n\alpha_n \bar{c}_{\mathscr{P}} \tag{93a}$$

$$\bar{D}_{H^+} = \left(D_{H^+}\frac{\partial c_{H^+}}{\partial x} + \sum_{n=0}^{2v} nD_n \frac{\partial(\alpha_n \bar{c}_{\mathscr{P}})}{\partial x}\right)\bigg/\left(\frac{\partial c_{H^+}}{\partial x} + \sum_{n=0}^{2v} n\frac{\partial(\alpha_n \bar{c}_{\mathscr{P}})}{\partial x}\right) \tag{93b}$$

$$\bar{\mu}_{H^+} = (c_{H^+}\mu_{H^+} + \sum_{n=0}^{2v} n\alpha_n \bar{c}_{\mathscr{P}}\mu_n)\bigg/(c_{H^+} + \sum_{n=0}^{2v} n\alpha_n \bar{c}_{\mathscr{P}}) \tag{93c}$$

To complete this analysis, it is necessary to relate the hydrogen ion concentration to the concentrations of buffer acid and its conjugate base by means of the usual equilibrium expression for the acid dissociation constant, and to express the electric field strength in terms of the current and the valences, concentrations, and mobilities of the macromolecular ions and the small ions of the supporting electrolyte (7). When this is done, Eqs. (87), (89), (92), and (93), together with the latter two expressions, describe completely the electrophoretic behavior of an inherently homogeneous protein under actual experimental situations. Solution of this set of equations gives not only the electrophoretic patterns of the protein, but also the gradients of field strength and pH established in the electrophoresis column by the electrophoretic process. Unfortunately, this is a formidable mathematical problem, except in the limit of ideal diffusion and electrophoresis of a system sufficiently well buffered that the pH is not disturbed significantly during transport of the protein. In that event, the quantities α_n, $\bar{D}_\mathscr{P}$, E, and $\bar{\mu}_\mathscr{P}$ do not vary through the protein boundary. Accordingly, Eq. (92) reduces to

$$\frac{\partial \bar{c}_\mathscr{P}}{\partial t} = \bar{D}_\mathscr{P} \frac{\partial^2 \bar{c}_\mathscr{P}}{\partial x^2} - \bar{\mu}_\mathscr{P} E \frac{\partial \bar{c}_\mathscr{P}}{\partial x} \tag{94}$$

This equation tells us that ideal ascending and descending electrophoretic patterns are enantiographic and show a single Gaussian-shaped boundary spreading with the constituent diffusion coefficient and migrating with the constituent mobility. Thus, in this limit only, an ampholyte exhibits the same behavior as a single macromolecular ion. It is also of interest to note the analogy to those isomerizing systems in which equilibrium is reestablished instantaneously during differential transport of the isomers. By simply incorporating the hydrogen ion activity into the acid dissociation constants of Eq. (87), the ideal amphoteric system can be reduced formally to a system of $2v + 1$ isomers.

In practice, it is difficult to realize conditions of ideality (see Chapter I), and theoretical prediction of the precise shape of experimentally observed boundaries and detailed description of the changes in pH and field strength across the boundaries must await numerical solution of the above equations. However, a considerable amount of information can be obtained from the conservation equation [Eq. (92)] without recourse to its solution. For example, examination of the moments of the boundary leads to definition of its position and insight into the factors governing its spreading with time.

The first moment of a moving boundary across which the constituent $\mathscr{P}$ disappears is given by

$$\bar{x} = \int_{x^\beta}^{x^\alpha} x \frac{\partial \bar{c}_\mathscr{P}}{\partial x} dx \bigg/ \int_{x^\beta}^{x^\alpha} \frac{\partial \bar{c}_\mathscr{P}}{\partial x} dx \tag{95}$$

where the limits of integration are reference positions in the two homogeneous solutions[1] (phases) on either side of the boundary (Fig. 13a). These positions

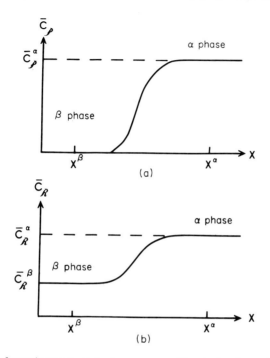

FIG. 13. Plot of constituent concentration versus position in the electrophoresis column: (a) the moving boundary formed upon electrophoresis of a solution containing a single protein constituent $\mathscr{P}$; (b) change in concentration of constituent $\mathscr{R}$ across a moving boundary corresponding to one of the other constituents of a mixture of proteins. Migration proceeds toward the right or left, depending upon whether the boundary is descending or ascending in the Tiselius cell.

are stationary with respect to the average water molecules and are sufficiently far apart that migration of the boundary will not bring a reference position into the region of changing concentrations of protein and small ions of the supporting electrolyte. Integration by parts gives

$$\bar{x} = x^\alpha - (1/\bar{c}_\mathscr{P}^\alpha) \int_{x^\beta}^{x^\alpha} \bar{c}_\mathscr{P}\, dx \qquad (96)$$

[1] Svensson (9) has made complete records of the conductivity throughout moving-boundary systems. It is evident that the center portion of the generated phase between the ε boundary and the descending moving boundary is indeed homogeneous with respect to salt ions as well as protein. The same is at least approximately so between the δ boundary and ascending boundary.

Upon differentiation with respect to time and substitution of Eq. (92) for $\partial \bar{c}_\mathscr{P}/\partial t$, we obtain

$$\frac{d\bar{x}}{dt} = -\frac{1}{\bar{c}_\mathscr{P}{}^\alpha}\int_{x\beta}^{x\alpha}\frac{\partial \bar{c}_\mathscr{P}}{\partial t}dx = -\frac{1}{\bar{c}_\mathscr{P}{}^\alpha}\int_{x\beta}^{x\alpha}\frac{\partial}{\partial x}\left[\bar{D}_\mathscr{P}\frac{\partial \bar{c}_\mathscr{P}}{\partial x} - \bar{\mu}_\mathscr{P}E\bar{c}_\mathscr{P}\right]dx \tag{97}$$

Integration yields the fundamental relationship

$$d\bar{x}/dt = \bar{\mu}_\mathscr{P}{}^\alpha E^\alpha \tag{98}$$

which tells us that the rate of migration of the first moment is the constituent velocity of the protein in the phase ahead of the descending or behind the ascending boundary. This is so, irrespective of the shape of the boundary and the presence of gradients of pH and field strength across the boundary. The rate of migration of the first moment of the descending boundary gives the value of the constituent mobility in the dialyzed protein solution (α phase) used to form the initial boundary, provided the electric field strength is computed from the conductance of that solution. In the case of the ascending boundary, $d\bar{x}/dt$ gives $\bar{\mu}_\mathscr{P}$ in the homogeneous phase generated between the δ boundary and the moving boundary when the conductance of that phase is used to compute the field strength. But, in general, this value of $\bar{\mu}_\mathscr{P}$ will differ from that in the original protein solution because of different pH.

While the rate of migration of a boundary is determined solely by the constituent velocity in the α phase, its spreading is governed by both random diffusion and variation in constituent velocity within the boundary. Thus, the time rate of change of the variance of the boundary is found to be

$$\frac{d(\overline{x^2} - \bar{x}^2)}{dt} = 2\bar{D}_\mathscr{P}{}^\alpha + \frac{2}{\bar{c}_\mathscr{P}{}^\alpha}\int_{x\beta}^{x\alpha}(\bar{\mu}_\mathscr{P}{}^\alpha E^\alpha - \bar{\mu}_\mathscr{P}E)\bar{c}_\mathscr{P}\,dx$$

$$-\frac{2}{\bar{c}_\mathscr{P}{}^\alpha}\int_{x\beta}^{x\alpha}\bar{c}_\mathscr{P}\frac{\partial \bar{D}_\mathscr{P}}{\partial x}dx \tag{99}$$

It is apparent that boundary spreading may be either greater or less than that expected from diffusion alone, although quantitative deductions must obviously await solution of the conservation equations. In the meantime, appeal can be made to the Dole strong-electrolyte moving-boundary theory (8). This theory permits estimation of ionic compositions of generated phases (in the notation of Fig. 13, these would be the α phase between the δ and ascending boundaries and the β phase between the ε and descending boundaries) from the known composition of the dialyzed protein and buffer solutions used to form the initial boundary. An illustrative example is provided by serum albumin in 0.1 ionic strength sodium diethylbarbiturate buffer, pH 8.6, a system in which both the protein and the conjugate buffer base migrate anodically. The Dole theory correctly predicts (7, 9, 10) that, for such a protein system, the electrical conductance of the α phase will be less than that of the β phase. Consequently, the electric field strength, and therefore the constituent velocity, will decrease on passing through the boundary from the α to the β phase. This being the case, Eq. (99) tells us that, whereas the spreading of the descending boundary will be greater than expected from diffusion alone, the spreading of the ascending boundary will be less than expected.[2] Mechanistically, the macromolecular ions in the trailing edge of the descending boundary experience a smaller electrical force, and thus migrate more slowly than those in the leading edge. This discordancy causes nonrandom broadening of the boundary. The opposite pertains to the ascending boundary, within which trailing macromolecular ions tend to catch up with leading ones. The resulting nonenantiography is illustrated by the electrophoretic patterns of serum albumin presented in Fig. 3.

Fundamental to the theory of moving boundaries is the moving boundary equation. Like the conservation equation from which it derives, the moving-boundary equation is simply a statement of conservation of mass during electrophoretic transport. Let us consider a moving boundary separating two phases, the β and α phases of Fig. 13b, both of which contain the constituent $\mathscr{R}$. An illustration would be the albumin boundary in the descending electrophoretic pattern of serum, Fig. 4. All of the globulins as well as the salt ions and buffer acid are present on either side of the albumin boundary. We might choose γ-globulin, for example, as the $\mathscr{R}$th constituent. Suppose that the boundary sweeps through the volume $V^{\alpha\beta}$ cm^3 during the time interval $\Delta t = t_2 - t_1$; we inquire as to the accumulation of $\mathscr{R}$ in the displaced volume. The change in constituent concentration of $\mathscr{R}$ at any position in the electrophoresis column during this time interval is

$$\int_{t_1}^{t_2} (\partial \bar{c}_{\mathscr{R}}/\partial t) \, dt \qquad (100)$$

so that the change in the amount of $\mathscr{R}$ in the volume element $\mathscr{A} \, dx$ is

$$\mathscr{A} \left(\int_{t_1}^{t_2} (\partial \bar{c}_{\mathscr{R}}/\partial t) \, dt \right) dx \qquad (101)$$

---

[2] The quantities $\bar{\mu}$, $\bar{c}$, $V^{\alpha\beta}$, and $v^{\alpha\beta}$ in Eqs. (95)–(109) are given signs: $\bar{\mu}$ and $\bar{c}$ are given the sign of the ions which are in equilibrium, while $V^{\alpha\beta}$ and $v^{\alpha\beta}$ are positive or negative depending on whether the boundary moves toward the cathode or anode. Accordingly, given the experimentally verified (8, 10) prediction of the Dole theory, the first integral on the right-hand side of Eq. (99) will have a positive or negative value depending on whether the boundary is descending or ascending. In general, the value of the second integral will be negligibly small.

Accordingly, the accumulation of $\mathscr{R}$ in $V^{\alpha\beta}$ is obtained by integrating across the boundary in the direction of migration

$$\mathscr{A} \int_{x^\beta}^{x^\alpha} \int_{t_1}^{t_2} (\partial \bar{c}_\mathscr{R}/\partial t) \, dx \, dt \tag{102}$$

By substituting the conservation equation (92) for $\partial \bar{c}_\mathscr{R}/\partial t$ and integrating with respect to both position and time, we find that the amount of $\mathscr{R}$ accumulated is equal to

$$\mathscr{A} \int_{x^\beta}^{x^\alpha} \int_{t_1}^{t_2} \frac{\partial}{\partial x}\left(\bar{D}_\mathscr{R} \frac{\partial \bar{c}_\mathscr{R}}{\partial x} - \bar{\mu}_\mathscr{R} E \bar{c}_\mathscr{R}\right) dx \, dt = -(\bar{\mu}_\mathscr{R}^\alpha E^\alpha \bar{c}_\mathscr{R}^\alpha - \bar{\mu}_\mathscr{R}^\beta E^\beta \bar{c}_\mathscr{R}^\beta) \mathscr{A} \, \Delta t \tag{103}$$

Since the increase in concentration of $\mathscr{R}$ across the boundary at all times is

$$\int_{x^\beta}^{x^\alpha} d\bar{c}_\mathscr{R} = \bar{c}_\mathscr{R}^\alpha - \bar{c}_\mathscr{R}^\beta \tag{104}$$

the accumulation can also be expressed as

$$-(\bar{c}_\mathscr{R}^\alpha - \bar{c}_\mathscr{R}^\beta) V^{\alpha\beta} \tag{105}$$

Conservation of mass requires that expressions (103) and (105) be equated:

$$(\bar{\mu}_\mathscr{R}^\alpha E^\alpha \bar{c}_\mathscr{R}^\alpha - \bar{\mu}_\mathscr{R}^\beta E^\beta \bar{c}_\mathscr{R}^\beta) \mathscr{A} \, \Delta t = V^{\alpha\beta}(\bar{c}_\mathscr{R}^\alpha - \bar{c}_\mathscr{R}^\beta) \tag{106}$$

Recalling that $E = I/\mathscr{H} \mathscr{A}$, where $\mathscr{H}$ is the specific conductance of the phase through which the electrical current I is flowing, Eq. (106) can be written as

$$\frac{\bar{\mu}_\mathscr{R}^\alpha \bar{c}_\mathscr{R}^\alpha}{\mathscr{H}^\alpha} - \frac{\bar{\mu}_\mathscr{R}^\beta \bar{c}_\mathscr{R}^\beta}{\mathscr{H}^\beta} = v^{\alpha\beta}(\bar{c}_\mathscr{R}^\alpha - \bar{c}_\mathscr{R}^\beta) \tag{107}$$

where $v^{\alpha\beta} = V^{\alpha\beta}/I \, \Delta t$ is the volume displaced by the moving boundary during passage of one coulomb of current. Equation (107) is the desired moving-boundary equation. It explicitly states that the net flow of $\mathscr{R}$ across two planes normal to the field, one located at x^α and the other at x^β, must equal the accumulation of $\mathscr{R}$ in the volume swept through by the boundary during the same time interval.

For the special case in which $\mathscr{R}$ is present only in the α phase, i.e., it disappears across the boundary, Eq. (107) tells us that

$$\bar{\mu}_\mathscr{R}^\alpha = v^{\alpha\beta} \mathscr{H}^\alpha \tag{108}$$

But, from Eq. (98),

$$\bar{\mu}_\mathscr{R}^\alpha = \frac{1}{E^\alpha} \frac{d\bar{x}}{dt} = \frac{d(\mathscr{A} \bar{x})}{d(It)} \mathscr{H}^\alpha \tag{109}$$

which reveals that the volume swept through by the boundary is defined by the displacement of the first moment of the boundary. The fact that the correct boundary position is the position of the first moment was first recognized by Longsworth (*11*).

The practical importance of the moving-boundary equation resides in the fact that it permits computation of the composition of a given phase in the electrophoresis column from the composition of the succeeding phase on the other side of either a moving or stationary (ε or δ) boundary. The change in concentration of constituent $\mathscr{R}$ across a moving boundary corresponding to some other constituent of the system is analogous in both origin and expression to the Johnston–Ogston effect in sedimentation. The analogy can be seen with greatest clarity by rearranging Eq. (107) to read

$$\frac{\bar{c}_{\mathscr{R}}^{\alpha}}{\bar{c}_{\mathscr{R}}^{\beta}} = \frac{\mu^{\alpha\beta} - \bar{\mu}_{\mathscr{R}}^{\beta}\,(\mathscr{H}^{\alpha}/\mathscr{H}^{\beta})}{\mu^{\alpha\beta} - \bar{\mu}_{\mathscr{R}}^{\alpha}} \tag{110}$$

where $\mu^{\alpha\beta} = v^{\alpha\beta}\mathscr{H}^{\alpha}$, and comparing with Eq. (53), which describes the Johnston–Ogston effect. (The factor $\mathscr{H}^{\alpha}/\mathscr{H}^{\beta}$ makes correction for the change in electric field across the $\alpha\beta$ boundary.) For a stationary boundary, $v^{\alpha\beta} = 0$, and

$$\frac{\bar{c}_{\mathscr{R}}^{\alpha}}{\bar{c}_{\mathscr{R}}^{\beta}} = \frac{\bar{\mu}_{\mathscr{R}}^{\beta}}{\bar{\mu}_{\mathscr{R}}^{\alpha}}\frac{\mathscr{H}^{\alpha}}{\mathscr{H}^{\beta}} = \frac{\bar{\mu}_{\mathscr{R}}^{\beta}}{\bar{\mu}_{\mathscr{R}}^{\alpha}}\frac{E^{\beta}}{E^{\alpha}} \tag{111}$$

This equation shows quite clearly that the ε and δ boundaries *per se* are expressions of conservation of mass during electrophoretic transport. Thus, for a stationary boundary, the net flow of $\mathscr{R}$ across the plane located at x^{α} exactly equals the net flow of $\mathscr{R}$ across the plane located at x^{β}, i.e., $\bar{c}_{\mathscr{R}}^{\alpha}\bar{\mu}_{\mathscr{R}}^{\alpha}E^{\alpha} = \bar{c}_{\mathscr{R}}^{\beta}\bar{\mu}_{\mathscr{R}}^{\beta}E^{\beta}$.

Recognition of the fact that the departures of both electrophoretic and ultracentrifugal transport from ideal behavior have a common cause, namely, the conservation of mass, brings a unity to the field which might otherwise go unnoticed. Both are forced-diffusion processes subject to the same fundamental laws of mechanics, the only difference being that the external force acting on the molecules in the one case is a real electrical force, and in the other case is a noninertial centrifugal force.

Above, we treated the electrophoresis of a protein from the point of view of the transport of a reversibly reacting system in which one of the reactants is the hydrogen ion; the protein boundary was regarded as a reaction boundary within which acid–base equilibria are continually readjusting during differential transport of the macroions PH_n. The concepts elaborated constitute what is known as the weak-electrolyte moving-boundary theory. They are equally applicable to other protein–small ion interactions such as enzyme–effector, serum albumin–methyl orange (*12–14*), serum albumin–chloride ion

(15), and serum albumin–Cd^{2+} (16). It is interesting to note that, if the protein interacts with a small, electrically neutral molecule, e.g., undissociated buffer acid, the neutral molecule can be treated formally as an ion constituent of the system, since it is transported electrophoretically in the form of protein complexes. [See, however, Cann and Goad (17, footnote 7) for discussion of the inherent and serious limitations on the application of the weak-electrolyte moving-boundary theory to protein–buffer acid interactions.] The theory also provides a means for quantitative interpretation of the electrophoretic behavior of certain types of systems in which different macromolecules interact with each other. Examples include the electrophoretic demonstration of the specific Michaelis–Menten complex between pepsin and serum albumin (18–20), complex formation between ovalbumin and nucleic acid (7), between conalbumin and lysozyme (21), and insulin–protamine (22). Application of the weak-electrolyte moving-boundary theory to the quantitative characterization of representative, rapidly equilibrating macromolecular interactions is illustrated below.

INTERACTIONS OF PROTEINS WITH SMALL IONS

As pointed out in Chapter I, the mobility of a protein is dependent not only on pH and ionic strength, but also buffer composition. This is so because the net charge on the protein molecule is determined by both the state of ionization of its neutral and cationic acid groups and the binding of buffer ions other than hydrogen ion. Moreover, binding of salt ions is often competitive in nature; and the bound ions affect the acid–base equilibria. In addition, complex formation with undissociated buffer acid frequently changes the net charge of the protein. It is apparent that a protein can participate in a large number of equilibria and that the number of macromolecular ionic species in solution may be much greater than expected from the acid–base equilibria alone. For example, the state of a protein in sodium acetate–acetic acid–sodium chloride (NaAc–HAc–NaCl) buffer might be represented by the constellation of macroions, $PH_i{}^+ Ac_j{}^- Cl_k{}^- (HAc)_l$, each of which contributes to the constituent mobility of the protein. Accordingly, changes in the value of the constituent mobility resulting from changes in buffer composition may be difficult to interpret. Consider serum albumin in NaAc–HAc buffer, pH 5.5. Substitution, at constant pH and ionic strength, of a portion of the NaAc with NaCl produces a significant change in constituent mobility (23). However, estimation of the number of Cl^- bound by the protein from the change in its mobility may be complicated by concomitant changes in the numbers of Ac^-, H^+, and HAc that are bound. In their experiments on the binding of Cl^- by serum albumin, Alberty and Marvin (15) circumvented these difficulties by (1) dispensing with the buffer entirely and relying on the buffer

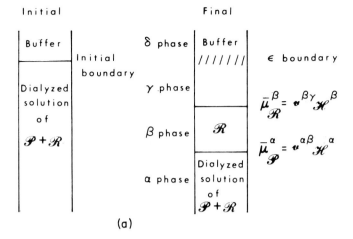

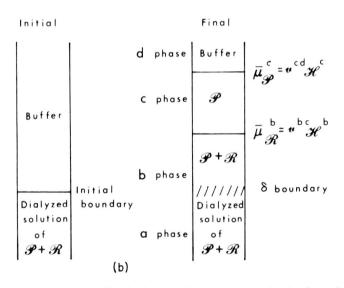

Fig. 14. Notation used to identify the several phases present in the descending and ascending limbs of the U-shaped Tiselius electrophoresis cell after resolution of the moving boundaries. Illustrated for a mixture of two constituent proteins $\mathscr{P}$ and $\mathscr{R}$ which do not interact with each other: (a) descending limb; (b) ascending limb. The boundary separating the α and β phases is designated as the $\alpha\beta$ boundary, etc. The ions of the buffer salt and other supporting electrolytes exist throughout the moving-boundary system, but note that (1) their concentrations in the β phase, and, thus, the pH and conductance of this phase, differ from the α phase; (2) their concentrations in the γ phase differ from the β and δ phases; and (3) the constituent concentration of $\mathscr{R}$ changes across the $\alpha\beta$ boundary. The concentrations of small ions in the four ascending phases also differ; the constituent concentration of $\mathscr{P}$ changes across the bc boundary, and the constituent concentrations of both $\mathscr{P}$ and $\mathscr{R}$ change across the δ boundary. See Longsworth (7) and Cann (10) for method of estimating phase composition from the Dole theory.

capacity of the protein to control the pH, and (2) measuring the constituent mobilities of both Cl^- and serum albumin and analyzing these data with aid of the weak-electrolyte moving-boundary theory. Although this is the preferred procedure for the study of small-ion binding, experiments with unbuffered solutions are relatively difficult to perform. It is significant, therefore, that Smith and Briggs (12) have shown that essentially the same quantitative information concerning the interaction of serum albumin with methyl orange may be obtained by moving-boundary electrophoresis in acetate buffer as by the equilibrium dialysis method of Klotz et al. (24). Since this system has come to serve as a model for both electrophoretic and ultracentrifugal investigations of protein–small-ion interactions, it will be analyzed in detail.

Before proceeding, however, a few words concerning notation are in order. We adopt the notation of Longsworth for designating the several phases present in the Tiselius cell after resolution of the moving boundaries. The phases in the descending limb are identified by lower case Greek letters in order from the lowermost, dialyzed protein solution upward to and including the buffer solution. The phases in the ascending limb are identified by lower case Roman letters in order from the dialyzed protein solution upward to the buffer. This notation is illustrated in Fig. 14 for a mixture of two constituent proteins $\mathscr{P}$ and $\mathscr{R}$ which do not interact with each other. Here, the constituent notation takes cognizance of both acid–base equilibria and binding of salt ions. However, in applying the weak-electrolyte moving-boundary theory to particular interactions of proteins with small ions other than H^+, or with each other, the acid–base equilibria usually are ignored for pragmatic reasons[3]; the constituent notation is reserved for description of the particular interaction under examination. In other words, the uncomplexed protein molecule and each of its complexes with either a small ion or some other protein molecule are regarded as single macromolecular ions.

Mass-Action Considerations. Several assumptions (13) can be made regarding the nature of the interaction of a protein with a small ion A. These include: (1) all of the protein is present as a single complex; (2) some of the protein exists in an uncomplexed form which is in equilibrium with a single complex; and (3) the uncomplexed form of the protein P is in equilibrium with a number of complexes $PA, \ldots, PA_i, \ldots, PA_n$ whose concentrations follow a statistical distribution. Since Klotz (24) has shown that the binding of small ions by proteins may often be represented by the Langmuir adsorption isotherm, the third assumption makes the best approximation. If each protein molecule

---

[3] This simplification is obviously not possible in case of metal ions which compete with H^+ for the binding sites (16).

possesses a maximum of n binding sites, formation of the successive complexes may be represented by the following equilibria (13, 24):

$$P + A \rightleftharpoons PA; \quad K_1 = c_1/c_P c_A$$
$$\vdots$$
$$PA_{i-1} + A \rightleftharpoons PA_i; \quad K_i = c_i/c_{i-1} c_A \quad (112)$$
$$\vdots$$
$$PA_{n-1} + A \rightleftharpoons PA_n; \quad K_n = c_n/c_{n-1} c_A$$

where the subscripts P, A, and i indicate uncomplexed protein, unbound A, and the complex PA_i, respectively. If it is assumed that all of the binding sites are identical in their intrinsic affinity for A, then every protein–A bond is formed by an association reaction

$$\text{isolated binding site} + A \rightleftharpoons \text{isolated protein–A bond} \quad (113)$$

which is governed by the intrinsic equilibrium or association constant K_0 defined by the relationship

$$K_0 = (\text{isolated protein–A bond})/(\text{isolated binding site}) c_A \quad (114)$$

The association reaction (113) is equivalent to the reaction of A with a hypothetical univalent protein. If it is assumed further that there are no interactions between the binding of one ion and the binding of additional ones, if any, to the same protein molecule, the equilibrium constant of each of the equilibria (112) is given by the product of K_0 and an appropriate statistical factor,

$$K_i = \frac{n-i+1}{i} K_0 \quad (115)$$

With these assumptions, application of the law of mass action and Hess's law of constant heat summation to the equilibria yields

$$c_i = c_P c_A^i \binom{n}{i} K_0^i \quad (116)$$

for the concentration of each complex;

$$c_P = \bar{c}_{\mathscr{P}}/[1 + K_0 c_A]^n \quad (117)$$

for the concentration of uncomplexed protein; and the following relationship[4] between the moles of A bound per mole of total protein, $\bar{i}$, and the equilibrium concentration of unbound A:

$$\bar{i} = \frac{n K_0 c_A}{1 + K_0 c_A} \quad (118)$$

---

[4] Here, we have assumed ideal solutions. For nonideal solutions, $1/\bar{i}$ may vary with protein concentration, and plots of $1/\bar{i}$ vs. $1/c_A$ will depart from linearity (25, 29).

which can be rearranged to give

$$\frac{1}{\bar{\imath}} = \frac{1}{nK_0 c_A} + \frac{1}{n} \tag{119}$$

By definition,

$$\bar{\imath} = (\bar{c}_{\mathscr{A}} - c_A)/\bar{c}_{\mathscr{P}} \tag{120}$$

where the constituent concentrations of protein and interacting ion are given by

$$\bar{c}_{\mathscr{P}} = c_P + \sum_{i=1}^{n} c_i \tag{121}$$

$$\bar{c}_{\mathscr{A}} = c_A + \sum_{i=1}^{n} i c_i \tag{122}$$

Equation (119) provides a linear relationship which may be useful in the evaluation of the constants n and K_0. A plot of $1/\bar{\imath}$ vs. $1/c_A$ extrapolates to $1/n$ at zero value of $1/c_A$, i.e., at infinitely high concentration of interacting ion. The slope of the straight line gives the intrinsic association constant.

In the equilibrium dialysis method, an aliquot of a buffered salt solution containing a known concentration of protein is equilibrated by dialysis against a given volume of an A solution of known composition. After equilibration of A across the dialyzing membrane, the two solutions are analyzed for A and, if necessary, for protein. The constituent concentrations of protein and A in the equilibrium mixture and the equilibrium concentration of unbound A are thereby determined; and $\bar{\imath}$ is computed from Eq. (120). By systematically varying the initial concentration of A, $\bar{\imath}$ is obtained as a function of c_A. If the above assumptions as to the nature of the interaction apply to the particular system being investigated, n and K_0 can be evaluated from a plot of $1/\bar{\imath}$ vs. $1/c_A$. Otherwise, a more detailed analysis is required (*25–27, 28c*). The binding of methyl orange by bovine serum albumin evidently satisfies the assumptions. In their classical experiments on this system, Klotz *et al.* (*24*) found $n = 22.4$ and $K_0 = 2.23 \times 10^3$ liters mole^{-1}. In contrast to the binding of methyl orange, human serum albumin has two classes of binding sites differing in affinity for fatty acid anions (*28*). A more striking example of heterogeneity is afforded by the specific combination of a hapten with its highly purified antihapten antibody (*26, 27*). Although each antibody molecule has only two specific combining sites, the heterogeneity of the sites among the total population of antibody molecules is often quite severe. Indeed, in certain instances, it becomes necessary to interpret binding data in terms of a continuous distribution of sites with respect to their free energy of combination with hapten. Clearly, elucidation of the macromolecular structural source of this heterogeneity is important for understanding the mechanism of antibody formation.

An unusual type of binding behavior of considerable biological significance is exhibited by regulatory enzymes such as aspartate transcarbamylase (*28a–c*), which shows cooperative binding of substrate (and substrate analog) molecules at topographically distinct binding sites. This cooperativity is mediated by the enzyme molecule itself through changes in macromolecular conformation. Such allosteric interactions are expressed by a sigmoid, rather than a Langmuir-type, saturation curve of the enzyme by the ligand.

Above, we have emphasized characterization of protein–small-ion interactions in terms of the number and intrinsic affinity of combining sites. Considerable effort has also been devoted to determination of the chemical nature of the sites and the type of forces responsible for the interactions (*29*). A variety of physical methods in addition to equilibrium dialysis have been brought to bear on these problems. They include gel-filtration, electromotive-force, fluorescence, optical-rotation, sedimentation, and electrophoretic measurements.

Electrophoretic Measurements. It is evident from the foregoing considerations that characterization of an interacting system of known overall composition (i.e., known $\bar{c}_{\mathscr{P}}$ and $\bar{c}_{\mathscr{A}}$) requires the experimental determination of only a single unknown quantity, namely, c_A. Electrophoretic analysis permits deduction of c_A from measured mobilities, provided, of course, that the rates of complex formation and dissociation are fast compared with the duration of the experiment. We proceed as follows (*13*).

The constituent mobilities of the protein and the interacting ion are

$$\bar{\mu}_{\mathscr{P}} = (1/\bar{c}_{\mathscr{P}})(c_P \mu_P + \sum_{i=1}^{n} c_i \mu_i) \tag{123}$$

$$\bar{\mu}_{\mathscr{A}} = (1/\bar{c}_{\mathscr{A}})(c_A \mu_A + \sum_{i=1}^{n} i c_i \mu_i) \tag{124}$$

Substitution of Eq. (116) for the concentration of each complex into Eq. (124) gives

$$\bar{\mu}_{\mathscr{A}} = (1/\bar{c}_{\mathscr{A}})(c_A \mu_A + c_P \sum_{i=1}^{n} i c_A{}^i \binom{n}{i} K_0{}^i \mu_i) \tag{125}$$

which cannot be cast in closed form without information as to how μ_i depends upon the structure of the complex. Let us assume that binding of small ions, while changing the net charge on the macromolecule, has a negligible effect on its frictional coefficient. In that event, one might expect that

$$\mu_i = \mu_P + iw \tag{126}$$

where w is an empirical constant. By substituting this expression into Eq. (123) and summing, we find for the constituent mobility of the protein

$$\bar{\mu}_\mathscr{P} = \mu_P + \bar{i}w \tag{127}$$

Combination of Eqs. (117), (118), (120), and (125)–(127) yields an explicit relationship between c_A, the overall composition of the system, and experimentally measurable mobilities. We proceed by substituting Eq. (126) into Eq. (125), performing the two summations and simplifying with the aid of Eqs. (117) and (118). These operations yield

$$\bar{\mu}_\mathscr{A} = \frac{1}{\bar{c}_\mathscr{A}} \left\{ c_A \mu_A + \bar{c}_\mathscr{P} \bar{i} \left[\mu_P + w\left(1 + \frac{(n-1)}{n}\bar{i}\right) \right] \right\} \tag{128}$$

By eliminating w between Eqs. (127) and (128), recalling Eq. (120) and rearranging, we obtain the following expression for c_A:

$$c_A = \frac{\bar{c}_\mathscr{A}(\bar{\mu}_\mathscr{A} - \bar{\mu}_\mathscr{P}) - \bar{c}_\mathscr{P}(\bar{\mu}_\mathscr{P} - \mu_P)}{(\mu_A - \bar{\mu}_\mathscr{P})} \tag{129}$$

provided $(n-1)/n$ is sufficiently close to unity. The values of $\bar{c}_\mathscr{P}$ and $\bar{c}_\mathscr{A}$ are fixed by the amounts of protein and interacting ion used to prepare the reaction mixture, and μ_P and μ_A are taken as the mobilities of the protein and A in solutions of the pure constituents. It only remains to design moving-boundary electrophoretic experiments for determination of the constituent mobilities.

For brevity, we shall consider only those systems for which $|\mu_A| > |\mu_P|$ and $|\mu_i| > |\mu_P|$. Serum albumin–methyl orange at pH alkaline to the isoelectric point is such a system. Two types of experiments are indicated (13). In experiment I, illustrated diagrammatically in Fig. 15, the initial boundary is formed between (1) an underlying undialyzed buffer solution containing known concentrations of protein and interacting ion, and (2) buffer of the same pH and ionic strength. From Fig. 15, it is seen that A disappears across the descending $\alpha\beta$ boundary, which is a reaction boundary within which the equilibria (112) are continually adjusting during differential transport of the complexes PA_i. According to the moving-boundary equation [Eq. (107)], the rate of migration of this boundary gives the constituent mobility of A in the initial equilibrium mixture:

$$\bar{\mu}_\mathscr{A}{}^\alpha = v^{\alpha\beta} \mathscr{H}^\alpha \tag{130}$$

where $\mathscr{H}^\alpha$ is the specific conductance of the initial equilibrium mixture. Similarly, the ascending bc boundary is a reaction boundary which yields the constituent mobility of the protein in the b phase:

$$\bar{\mu}_\mathscr{P}{}^b = v^{bc} \mathscr{H}^b \tag{131}$$

However, due to dilution across the δ boundary, the concentrations of reactants in the b phase are less than in the initial equilibrium mixture. Moreover, the required specific conductance $\mathscr{H}^b$ is difficult to measure. A more accurate

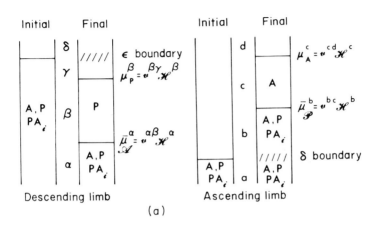

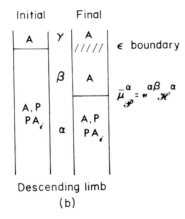

FIG. 15. Electrophoretic moving-boundary systems designed to characterize protein–small-ion interactions in which $|\mu_A| > |\mu_P|$ and $|\mu_i| > |\mu_P|$: (a) experiment I; (b) experiment II. See text for details. The ions of the buffer exist throughout the moving-boundary system and are not indicated.

method for determination of the constituent mobility of the protein is provided by experiment II of Fig. 15, in which A is present throughout the system of boundaries. In this experiment, the initial boundary is formed between (1) an underlying buffer solution of protein and A, and (2) buffer containing A. Because the concentration of A in the β phase will be somewhat larger than in the γ phase, a preliminary experiment or two may be required to determine the correct concentration of A to be used in the buffer so that $c_A{}^\beta = c_A{}^\alpha$. Having taken this precaution, the descending $\alpha\beta$ boundary gives the constituent mobility of the protein in the initial equilibrium mixture:

$$\bar{\mu}_\mathscr{P}{}^\alpha = v^{\alpha\beta}\,\mathscr{H}^\alpha \tag{132}$$

Note that this boundary is also a reaction boundary.

In their study of the binding of methyl orange by bovine serum albumin, Smith and Briggs (12) performed two different sets of experiments. Those of the first set were designed after experiment II for determination of $\bar{\mu}_\mathscr{P}$ and were coupled with equilibrium dialysis measurements. The results of these experiments are presented in Fig. 16. They clearly establish that $\bar{\mu}_\mathscr{P}$ is a linear

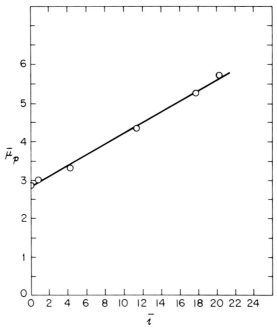

FIG. 16. Linear dependence of the constituent mobility of serum albumin $\bar{\mu}_\mathscr{P}$ (cm^2 sec^{-1} V^{-1}) upon the moles of methyl orange bound per mole of total albumin $\bar{\imath}$. Experiments designed after experiment II of Fig. 15. [From Smith and Briggs (12).]

function of $\bar{\imath}$, as assumed in the above derivation of Eq. (129) relating the equilibrium concentration of unbound A to overall composition and electrophoretic mobilities.

In the second set of experiments, albumin–methyl orange mixtures containing the same $\bar{c}_{\mathscr{P}}$ but progressively increasing $\bar{c}_{\mathscr{A}}$ were analyzed according to the design of experiment I. The resulting electrophoretic patterns are displayed in Fig. 17. The fast-migrating descending boundary is the $\alpha\beta$ (reaction)

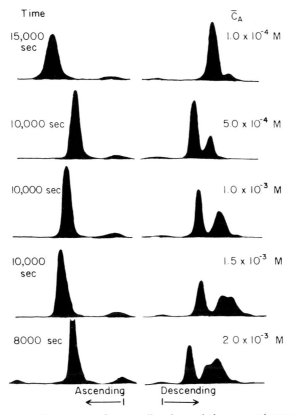

FIG. 17. Electrophoretic patterns of serum albumin–methyl orange mixtures containing constant $\bar{C}_{\mathscr{P}}$ but progressively increasing $\bar{C}_{\mathscr{A}}$. Here, $\bar{C}_{\mathscr{P}} = 1.14 \times 10^{-4} M$; values of $\bar{C}_{\mathscr{A}}$ shown; 0.05 M sodium acetate buffer, pH 5.5; time of electrophoresis shown; electric field strength, about 7 V cm^{-1}. Experiments designed after experiment I of Fig. 15. [From Smith and Briggs (*12*).]

boundary, and the slower one is the $\beta\gamma$ boundary corresponding to uncomplexed protein. The ascending boundary is the bc (reaction) boundary, the faster cd boundary corresponding to uncomplexed methyl orange being evident only in the pattern shown for the highest value of $\bar{c}_{\mathscr{A}}$. It is interesting that the $\alpha\beta$ boundary is bimodal for $\bar{C}_{\mathscr{A}} \geq 1.5 \times 10^{-3} M$ (C denotes molar concentration). Smith and Briggs present equilibrium dialysis measurements which suggest that serum albumin may have two classes of binding sites differing in their affinity for methyl orange, in which case, the bimodality may reflect binding at the weaker sites. Accordingly, only those patterns for which $\bar{C}_{\mathscr{A}} \leq 10^{-3} M$ were used to determine n and K_0.

In their analysis of the patterns, Smith and Briggs used a combination of area and mobility measurements to deduce the equilibrium concentration of unbound methyl orange with the aid of certain relationships derived in their paper. For this purpose, $\bar{\mu}_{\mathscr{A}}$ was calculated from the rate of migration of the $\alpha\beta$ boundary; μ_P from the $\beta\gamma$ boundary using $\mathscr{H}^\alpha$; and $\bar{\mu}_{\mathscr{P}}$ from the bc boundary using $\mathscr{H}^\alpha$ corrected for dilution across the δ boundary. Values of $\bar{\imath}$ were then obtained from Eq. (120), and the data analyzed graphically in accordance with Eq. (119). Although their values of n and K_0 are in satisfactory agreement with those of Klotz and co-workers (24) from equilibrium dialysis measurements, the author prefers the more accurate procedure described above for calculating C_A. By means of Eq. (129), C_A can be deduced without resort to area measurements. This has been done using the values of $\bar{C}_{\mathscr{P}}$, $\bar{C}_{\mathscr{A}}$, $\bar{\mu}_{\mathscr{P}}$, and $\bar{\mu}_{\mathscr{A}}$ given in Table 1 of Smith and Briggs and their values for μ_P and μ_A in solutions of pure albumin and methyl orange. The results are presented graphically in Fig. 18. The values of n and K_0 are in reasonable agreement

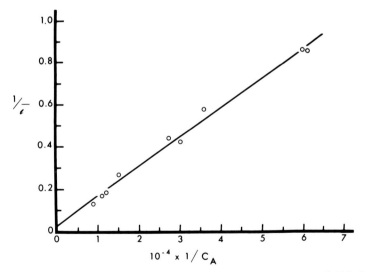

FIG. 18. Quantitative interpretation of electrophoretic measurements on the binding of methyl orange by serum albumin. In accordance with Eq. (119), the reciprocal of the number of moles of methyl orange bound per mole of total protein, $1/\bar{\imath}$, is plotted against the reciprocal of the equilibrium concentration of unbound dye, $1/C_A$. The least-square straight line gives $n = 32$ and $K_0 = 2.3 \times 10^3 \, M^{-1}$. See text for details concerning the calculation of C_A and $\bar{\imath}$ from overall composition of albumin–methyl orange mixtures and electrophoretic mobilities.

with those from equilibrium dialysis measurements considering (1) the scatter of the data points, (2) the fact that electrophoretic experiments were made in an acetate rather than the phosphate buffer used by Klotz and his co-workers, and (3) the fact that $\bar{\mu}_\mathscr{P}$ had been calculated from the ascending patterns. It would have been preferable had $\bar{\mu}_\mathscr{P}$ been determined in companion experiments designed according to experiment II. In any case, it is clear that electrophotetic measurements permit quantitative characterization of protein–small-ion interactions in buffered solutions.

Since the pioneering experiments of Smith and Briggs, the concepts of the weak-electrolyte moving-boundary theory have been successfully applied of several other protein–small-ion interactions. Reference has already been made to the highly precise measurements of Alberty and Marvin (15) on the binding of Cl^- by serum albumin. Measurements at pH 7.00, 5.40, and 3.20 indicate the binding of 8, 9, and 29 chloride ions per molecule of albumin. In view of the importance of this particular interaction in determining the properties of serum albumin under physiological conditions, it is gratifying that these values are in excellent agreement with those obtained by other methods. In his study of the interaction of serum albumin with Cd^{2+}, Schilling (16) took into account the competition between Cd^{2+} and H^+ for the same binding sites on the protein molecule. The resulting intrinsic association constant for Cd^{2+} agreed with that found for binding to histidine residues. Such measurements are often important to the interpretation of independent physicochemical measurements. For example, Cann and Phelps (30, 31) found that under certain conditions (for example, ionic strength 0.1 NaCl, pH changing from 7.0 to 3.1), bovine γ-pseudoglobulin undergoes conformational changes which affect its sedimentation constant but are without influence on its molecular weight. With properly chosen concentration and type of salt, however, acid pH's also can cause aggregation of the protein. Sedimentation experiments showed that the effectiveness of univalent anions in aggregating the protein at pH 3.1 increases in the order Cl^-, Br^-, NO_3^-, and ClO_4^-. Electrophoretic measurements revealed that the mean isoelectric point of the protein decreases on substitution of Br^-, NO_3^-, and ClO_4^- for Cl^- in the buffer solvent, the effectiveness of the substituting anions increasing in the mentioned order. The correlation between the effects of the various anions on the isoelectric pH and the extent of the aggregation of the protein at pH 3.1 is striking. Since one would expect that the more strongly bound a given anion, the more effective it will be in decreasing the isoelectric pH, these data indicate that differences in binding of anions, presumably, by the positively charged amino groups of the protein, afford an explanation of the differences in the sedimentation behavior of γ-globulin solutions of different supporting electrolytes at pH 3.1.

Finally, mention should be made of crossing-paper electrophoresis, an adaptation of conventional paper electrophoresis to the detection of reversible interactions (32, 33). The principle involved is that two oblique or perpendicular lines of different substances moving in an electric field will show a deformation at their point of crossing if they interact, because the complex will have a different mobility than at least one of the reactants. It is a rapid method for screening a large number of substances for possible interaction with a given protein and has revealed definite small-molecule structural requirements for complexing of amines with serum albumin (34).

Sedimentation Measurements. In view of the marked differences between the sedimentation coefficients of small and large molecules, sedimentation velocity holds promise as a powerful means for determining the extent of binding of small molecules and ions by macromolecules. It is surprising, therefore, that the sedimentation-velocity method has enjoyed only limited application to

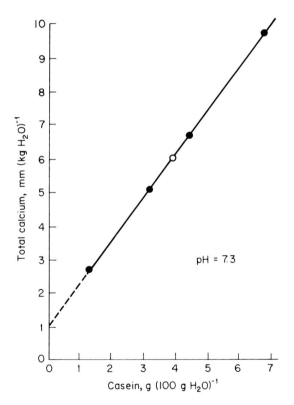

FIG. 19. Distribution of total calcium ion concentration vs. casein concentration in preparative ultracentrifuge cell after 2 hr of centrifugation at 60,000 rpm. [From Chanutin *et al.* (35).]

the study of macromolecule–small-molecule interactions. This, despite the important results obtained in the few studies reported. Apparently, the first application was by Chanutin, Ludewig, and Masket (*35*), who employed a preparative ultracentrifuge to measure the binding of Ca^{2+} to casein. After several hours of high-speed centrifugation of a mixture of known $\bar{c}_{Ca^{2+}}$ and $\bar{c}_{\mathscr{P}}$, the contents of the centrifuge cell were separated into a series of fractions corresponding to increasing distance from the center of rotation. Analysis of the fractions provided the data for a plot (Fig. 19) of Ca^{2+} concentration against casein concentration along the tube, thereby permitting linear extrapolation to zero protein concentration. The corresponding concentration of Ca^{2+} was assumed to represent $c_{Ca^{2+}}$ in the initial equilibrium mixture. A set of such experiments were made for different $\bar{c}_{Ca^{2+}}$ at constant $\bar{c}_{\mathscr{P}}$. From the results, Klotz (*36*) derived $\bar{\imath}$ as a function of $C_{Ca^{2+}}$ and determined n and K_0 graphically (Fig. 20). This was evidently the first application of Eq. (119) to protein interactions other than acid–base equilibria. A somewhat analogous procedure was used by Velick and his co-workers (*37–39*) for measuring the binding of diphosphopyridine nucleotide to glyceraldehyde-3-phosphate dehydrogenase.

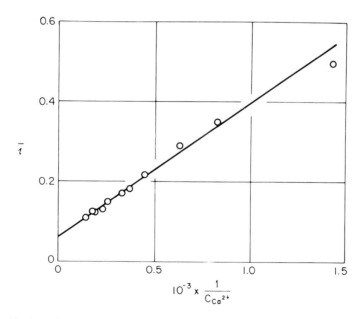

FIG. 20. Quantitative interpretation of sedimentation measurements on the binding of calcium ions by casein. Plot of the reciprocal of the number of moles of Ca^{2+} bound per mole of casein, $1/\bar{\imath}$, vs. the reciprocal of the equilibrium concentration of unbound Ca^{2+}, $1/C_{Ca^{2+}}$. Application of Eq. (119) gives $n = 16$ and $K_0 = 1.7 \times 10^2\ M^{-1}$. [From Klotz (*36*).]

Recently, zone sedimentation of macromolecule–small-molecule systems through density gradients has been used to great advantage by Gilbert and Müller-Hill (40, 41) in studies on the mechanism of enzyme induction and genetic control of carbohydrate metabolism in E. coli, in particular, the metabolism of lactose. The current state of knowledge of the lac system can be summarized as follows: The metabolism of lactose involves three specific proteins—the galactoside permease, which permits accumulation of β-galactosides in the cell; β-galactosidase, which hydrolyzes lactose to glucose and galactose; and a transacetylase of unknown function—whose synthesis is under the genetic control of the lac operon. The genes of the lac operon are coordinately and reversibly turned off by a repressor protein, which process constitutes a metabolic control. There is evidence that the repressor exerts its action by combining with a segment of DNA (the operator) adjacent to the genes controlled, thereby preventing their expression. Addition to a bacterial suspension of a small inducer molecule, either lactose itself or an analog such as isopropyl-1-thio-β-D-galactopyranoside (IPTG), causes the biosynthesis of the proteins of the lac operon, evidently by reacting reversibly with the repressor, thereby removing it from the operator. Direct evidence for repressor binding of IPTG was obtained by equilibrium dialysis, which also constituted the assay for identification and purification of the repressor protein. Now, in their studies of these processes, Gilbert and Müller-Hill employed two types of sedimentation experiments. In one type of experiment, the inducer, IPTG, was present in both the stabilizing gradient and the lac repressor fraction to be analyzed. Binding of IPTG was used as a device to locate the repressor in the centrifuge tube, thereby permitting estimation of its sedimentation coefficient, and, thus, its molecular weight. In the other experiment, it was demonstrated that, whereas in the absence of inducer, purified lac repressor binds specifically to the lac operator DNA, addition of inducer to the gradient solution prior to sedimentation of the macromolecular mixture causes dissociation of the repressor–operator complex.

The possibilities for analytical sedimentation analysis of interactions between large and small molecules have been tremendously increased by introduction of (1) the split-beam photoelectric scanning absorption system fitted with a monochromator for detecting and recording distributions of molecules in the ultracentrifuge cell, and (2) differential and comparative techniques of high precision for measuring small differences in sedimentation coefficients. Previously, the usefulness of the analytical ultracentrifuge for such analyses generally was restricted to situations in which interaction leads to a change in state of aggregation of the protein, e.g., insulin with thiocyanate (42) and mercaptalbumin with mercuric ion (43, 44). It is now possible in a single experiment to record the distributions of small, light-absorbing molecules using the photoelectric scanner and of large molecules using interference or Schlieren optics. In certain cases, the distributions of both small and large molecules can be recorded by photoelectric scanning at different wavelengths. The power of these methods as an aid in elucidating the mechanisms of biological reactions has been elegantly demonstrated by Schachman and his colleagues (28a–c) in their studies on allosteric interactions in the regulatory enzyme aspartate transcarbamylase (ATCase) from *E. coli*.

Aspartate transcarbamylase catalyzes the carbamylation of aspartate by carbamyl phosphate. This is the initial reaction unique to the biosynthesis of pyrimidines; and the enzyme is inhibited by cytidine triphosphate (CTP), an end product of the pyrimidine pathway, thus establishing in *E. coli* the regulatory pattern of feedback inhibition (see Refs. 28a–c for literature citations). It has been concluded from kinetic experiments that ATCase exhibits two classes of allosteric interactions[5] among the specific ligands involved in the regulation of its activity: (1) *homotropic* interactions, as evidenced by the sigmoid dependence of reaction velocity on the concentration of aspartate in the presence of saturating concentration of the other substrate, carbamyl phosphate, and (2) *heterotropic* interactions between the substrate (aspartate) and either the feedback inhibitor (CTP) or the activator (ATP). Both types of interaction have been demonstrated directly by equilibrium dialysis measurements on the binding of specific ligands to the native enzyme and to its isolated catalytic and regulatory subunits. *Homotropic* interactions were revealed by the cooperative binding of 3.8 ± 0.2 molecules of the substrate analog, succinate, per molecule of enzyme in the presence of saturating concentration of carbamyl phosphate; and *heterotropic* interactions by the antagonistic effect of CTP on succinate binding. Moreover, these interactions are lost when the ATCase molecule is dissociated into subunits, even though the catalytic subunit still binds the substrate analog and the regulatory subunit still binds the feedback inhibitor. The effect of ligands on the gross morphology of the intact enzyme was examined in the ultracentrifuge using a comparative sedimentation technique of high precision in which the two solutions to be compared are analyzed in a single experiment. In the presence of both succinate and carbamyl phosphate, there was a 3.6% reduction in the sedimentation coefficient of ATCase (Fig. 21). Both ligands were required, and the magnitude of the reduction was a function of the concentration of succinate at a fixed concentration of carbamyl phosphate (Fig. 22). Furthermore, the feed-

---

[5] Allosteric effects are defined as *indirect* interactions between topographically *distinct* binding sites mediated by the protein molecule through a conformational change (45, 46). *Homotropic* interactions occur between identical ligand molecules and are expressed by the sigmoidal saturation curve of the enzyme by the considered ligand. *Heterotropic* interactions occur between dissimilar ligand molecules and are typified by the effects of inhibitors and activators on the turnover number or apparent affinity (or both) of the enzyme for its substrate.

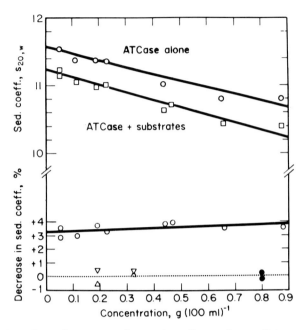

Fig. 21. Effect of protein concentration on the sedimentation coefficient at ATCase in the presence and absence of substrates. Samples were analyzed in pairs in a single ultracentrifuge experiment by examining the solution containing the substrates in one cell and the reference solution in a second cell. Upper graph is a plot of observed sedimentation coefficient against protein concentration: (○) ATCase in the reference solution containing glutamate and phosphate; (□) ATCase in solution containing the substrate analog, succinate, and carbamyl phosphate, pH 7.0. The lower graph shows the per cent decrease in sedimentation coefficient as measured in each single experiment with a pair of solutions: (○) calculated from experimental results presented in upper graph; (▽) control experiments on ATCase in presence of carbamyl phosphate alone; (△) controls in presence of succinate alone; (●) experiments on aldalase in presence of both carbamyl phosphate and succinate. [From Gerhart and Schachman (28b).]

back inhibitor, CTP, opposed the effect of succinate and carbamyl phosphate in reducing the sedimentation coefficient; and the antagonism was only partial. In the absence of other ligands, CTP had virtually no effect. No evidence for dissociation of the enzyme into subunits was obtained either from the appearance of the sedimenting boundaries or from the concentration dependence of the sedimentation coefficient. The observed changes in sedimentation coefficient appear to result from modification in the conformation of the undissociated enzyme molecule and to be the indirect effect of ligand molecules bound to the enzyme. In the presence of succinate and carbamyl phosphate, the intact enzyme molecule evidently exists in a swollen, less compact con-

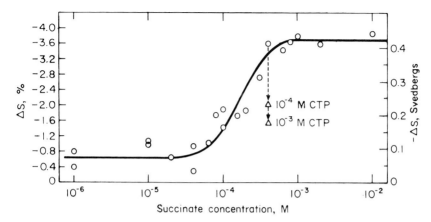

FIG. 22. Dependence of the sedimentation coefficient of ATCase on the concentration of the substrate analog, succinate, in the presence of $1.8 \times 10^{-3} \, M$ carbamyl phosphate, pH 7.0. Plot of change in sedimentation coefficient Δs against the succinate concentration in molarity: (○) in absence of CTP; (△) in presence of indicated concentrations of CTP. See Fig. 21 for experimental procedure. [From Gerhart and Schachman (28b).]

formation than in the absence of substrates. Binding of CTP appears to favor the more compact conformational state of ATCase. Thus, we see that the ATCase system satisfies the Monod–Wyman–Changeux criteria (45) for allosteric interactions.

In a related series of experiments, a combination of sedimentation-velocity and spectrophotometric measurements was used to investigate the effect of ligands on the reactivity of the sulfhydryl groups of ATCase as a chemical probe for changes in macromolecular conformation. Upon addition of p-mercuribenzoate (PMB) to form mercaptide complexes, the enzyme dissociates asymmetrically into two catalytic subunits and four regulatory subunits. Titration of ATCase with PMB, as measured by the dissociation of the enzyme into subunits, showed that complete dissociation requires 26 ± 2 molecules of PMB per molecule of ATCase, as compared with 27 ± 1 found for the reaction endpoint by spectrophotometric titration. In these experiments, Schlieren sedimentation patterns of ATCase–PMB mixtures were analyzed for undissociated enzyme. In companion experiments, sedimentation patterns of partially reacted mixtures were recorded at two different wavelengths using the photoelectric scanner: at 280 mμ, the light absorption was due almost entirely to the protein, whereas at 248 mμ, the absorption by the protein was approximately one-half its 280 mμ value and PMB absorbed strongly whether alone or as its mercaptide complex. Analysis of these patterns revealed that the undissociated enzyme bound no PMB, whereas the dissociated products were fully reacted. Moreover, in this *all-or-none* reaction of the sulfhydryl groups of the enzymes, virtually all of the PMB was bound to the regulatory subunits and practically none to the catalytic subunits. (Of the 32 half-cystines in the intact ATCase molecule as judged by cystic acid analysis, 24–28 are in the four regulatory subunits and 8 in the two catalytic subunits.) Although the rate of reaction of isolated regulatory subunits with PMB was too rapid for measurement by conventional spectrophotometry, the rate for the intact enzyme was readily determined. Addition of both succinate and carbamyl phosphate to the reaction mixture caused a sixfold increase in the pseudounimolecular rate of reaction. Succinate alone had no effect, and the effect of carbamyl phosphate alone was much less than that resulting from the addition of both ligands. Enhancement of the reactivity of sulfhydryl groups was opposed by the addition of CTP. This antagonism between substrates and feedback inhibitor was highly specific and only partial in character. In the absence of the substrates, CTP had almost no effect on the reactivity of the enzyme toward PMB. Evidently, the swollen conformational state of the enzyme in the presence of substrates is more reactive to PMB than the more compact conformation favored by the feedback inhibitor found in the absence of any ligands. The enhanced reactivity could conceivably be due to a greater rate of dissociation of the swollen protein molecule into subunits.

Above, we have summarized the results of the several applications of ultracentrifugation to the characterization of macromolecule–small-molecule interactions. It is apparent even from these few examples that sedimentation experiments can provide important information concerning the nature and biological significance of interactions between large and small molecules. Accordingly, it is noteworthy that Steinberg and Schachman (47) have made a detailed analysis of the sedimentation process itself in terms of the applicability and limitations of the methods of sedimentation velocity and sedimentation equilibrium for such systems. Theoretical formulations and experimental procedures are described and applied to a study of the interaction between bovine serum albumin and methyl orange. These are summarized as follows.

Velocity Sedimentation. The theory of velocity sedimentation of a rapidly equilibrating system of interacting constituents $\mathscr{P}$ and $\mathscr{A}$ which form the complexes PA_i in accordance with equilibria (112) is quite analogous to the weak-electrolyte moving-boundary theory of electrophoresis. Thus, the constituent sedimentation coefficients take the form

$$\bar{s}_{\mathscr{P}} = (1/\bar{c}_{\mathscr{P}})\left(c_P s_P + \sum_{i=1}^{n} c_i s_i\right) \tag{133}$$

$$\bar{s}_{\mathscr{A}} = (1/\bar{c}_{\mathscr{A}})\left(c_A s_A + \sum_{i=1}^{n} i c_i s_i\right) \tag{134}$$

If it is assumed that the interaction between P and each A molecule causes a constant increment m in the sedimentation coefficient of the macromolecule, the value for the complex PA_i can be written as

$$s_i = s_P + im \qquad (135)$$

and the constituent sedimentation coefficient of $\mathscr{P}$ becomes

$$\bar{s}_{\mathscr{P}} = s_P + \bar{i}m \qquad (136)$$

By following the same steps as in the derivation of Eq. (129), we obtain the analogous relationship

$$c_A = [\bar{c}_{\mathscr{A}}(\bar{s}_{\mathscr{P}} - \bar{s}_{\mathscr{A}}) + \bar{c}_{\mathscr{P}}(\bar{s}_{\mathscr{P}} - s_P)]/(\bar{s}_{\mathscr{P}} - s_A) \qquad (137)$$

For systems in which $\bar{s}_{\mathscr{P}} = s_P$, Eq. (137) reduces to

$$c_A = \bar{c}_{\mathscr{A}}(\bar{s}_{\mathscr{P}} - \bar{s}_{\mathscr{A}})/(\bar{s}_{\mathscr{P}} - s_A) \qquad (138)$$

The values of $\bar{c}_{\mathscr{P}}$ and $\bar{c}_{\mathscr{A}}$ are fixed by the quantities of the two constituents used to prepare the reaction mixture, and s_P and s_A can be taken as their sedimentation coefficients in pure solutions. It remains to determine the constituent sedimentation coefficients from the sedimentation pattern of the mixture.

As illustrated diagrammatically in Fig. 23, the sedimentation pattern will, in general, show two boundaries. The more rapidly sedimenting, $\beta\gamma$ boundary

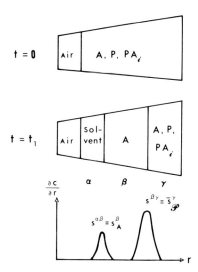

FIG. 23. Diagrammatic representation of the concentration distributions and the boundaries observed in the ultracentrifuge cell during velocity sedimentation of a rapidly equilibrating system of interacting constituents $\mathscr{P}$ and $\mathscr{A}$ which form the complexes PA_i. For the case, $s_P > s_A$. If A is a small molecule, such as methyl orange, its rate of sedimentation in the β phase will be so small that the $\alpha\beta$ boundary will not develop.

is a reaction boundary within which the equilibria (112) are continually readjusting during differential sedimentation of P and its complexes PA_i. The moving-boundary equation for sedimentation of noninteracting systems [Eq. (52)] also applies to interacting systems with only slight change in notation, and, for the $\mathscr{P}$ constituent, it states that

$$\bar{s}_{\mathscr{P}}^{\beta}\bar{c}_{\mathscr{P}}^{\beta} - \bar{s}_{\mathscr{P}}^{\gamma}\bar{c}_{\mathscr{P}}^{\gamma} = s^{\beta\gamma}(\bar{c}_{\mathscr{P}}^{\beta} - \bar{c}_{\mathscr{P}}^{\gamma}) \tag{139}$$

Since there is no protein in the β phase, Eq. (139) tells us that

$$s^{\beta\gamma} = \bar{s}_{\mathscr{P}}^{\gamma} \tag{140}$$

In other words, the sedimentation coefficient of the $\beta\gamma$ boundary is equal to the constituent sedimentation coefficient of $\mathscr{P}$ in the γ phase. Likewise, application of the moving-boundary equation to the slower-sedimenting $\alpha\beta$ boundary corresponding to pure A tells us that the sedimentation coefficient of this boundary is equal to the sedimentation coefficient of A in the β phase, $s^{\alpha\beta} = s_A^{\beta}$. In order to evaluate the constituent sedimentation coefficient of $\mathscr{A}$ in the γ phase, it is necessary to appeal once more to the moving-boundary equation for $\mathscr{A}$, but this time as applied to the $\beta\gamma$ boundary:

$$s_A^{\beta}c_A^{\beta} - \bar{s}_{\mathscr{A}}^{\gamma}\bar{c}_{\mathscr{A}}^{\gamma} = s^{\beta\gamma}(c_A^{\beta} - \bar{c}_{\mathscr{A}}^{\gamma}) \tag{141}$$

Upon rearrangement,

$$\bar{s}_{\mathscr{A}}^{\gamma} = s_A^{\beta}\frac{c_A^{\beta}}{\bar{c}_{\mathscr{A}}^{\gamma}} + \left(1 - \frac{c_A^{\beta}}{\bar{c}_{\mathscr{A}}^{\gamma}}\right)s^{\beta\gamma} \tag{142}$$

Absorption optical techniques are ideally suited for the implementation of Eq. (142), since c_A^{β} and $\bar{c}_{\mathscr{A}}^{\gamma}$ can be measured directly along with the necessary sedimentation coefficients. With Schlieren optics, the evaluation of $\bar{c}_{\mathscr{A}}^{\gamma}$ is not so direct, because the area of the $\beta\gamma$ boundary gives the value of $(\bar{c}_{\mathscr{A}}^{\gamma} + \bar{c}_{\mathscr{P}}^{\gamma} - c_A^{\beta})$; data from another experiment may be required for the determination of $\bar{c}_{\mathscr{P}}$. For systems in which A is a small molecule, such as methyl orange, the $\alpha\beta$ boundary will not develop. In that event, Eq. (142) is not applicable, and $\bar{s}_{\mathscr{A}}^{\gamma}$ must be evaluated from measurements of the transport of $\mathscr{A}$, both free and bound, across a surface located at position r_P in the γ phase using the transport equation

$$\bar{s}_{\mathscr{A}}^{\gamma} = (1/2\omega^2 t)\ln\{[(2/r_P^2\bar{c}_{\mathscr{A}})\int_{r_m}^{r_P} cr\,dr] + (r_m^2/r_P^2)\} \tag{143}$$

where c is determined from either photoelectric scanner patterns or conventional absorption patterns. The $\bar{s}_{\mathscr{A}}^{\gamma}$ is calculated from the slope of a plot of the logarithm of the quantity in brackets versus time. Thus, in principle, it is possible to evaluate, in a single sedimentation-velocity experiment, all of the quantities needed to compute c_A from Eq. (137) or (138) and $\bar{\imath}$ from Eq. (120).

A series of such experiments for constant $\bar{c}_{\mathscr{P}}$ but varying $\bar{c}_{\mathscr{A}}$ would provide the data required for characterization of the interaction with respect to n and K_0.

Before considering the practicality of applying the above formulations to actual situations, it is important to inquire as to the magnitude of the errors incurred if one simply equated c_A^{β} to c_A^{γ}. Upon ultracentrifugation of a protein–small-molecule system, the reaction boundary corresponding to the protein and its complexes sediments toward the bottom of the centrifuge cell, leaving behind unbound small molecules. One might expect the concentration of small molecule in the supernatant liquid to approximate its equilibrium concentration in the initial mixture. Depending upon the order of this approximation, it would become a relatively simple matter to quantitate interactions of large molecules with small ones. To test the validity of this procedure, Steinberg and Schachman performed sedimentation-velocity experiments on a series of serum albumin–methyl orange mixtures of known constituent concentrations. Sedimentation patterns were recorded with the photoelectric scanner using light of 334 mμ so as to read absorbances due only to methyl orange. Those obtained in one such experiment are displayed in Fig. 24. As seen in the first pattern, obtained shortly after the rotor had attained operating speed, the absorbance was virtually constant throughout the cell. After 32 min of centrifugation, the boundary due to the various complexes of protein with methyl orange, PA$_i$, had sedimented about one-third the distance through the cell. Left behind in the supernatant was unbound methyl orange. The progressive sedimentation of the boundary of the complexes is clearly evident in the subsequent patterns. The concentration of methyl orange at a fixed position in the supernatant was determined by comparison of recorder displacement with data from an additional experiment at the same constituent concentration of methyl orange but in the absence of protein. (Comparison with the initial pattern of the protein–methyl orange mixture is not appropriate, since free and bound methyl orange absorb differently at 334 mμ.) By assuming the supernatent concentration to be equal to the equilibrium concentration of unbound dye, values of $\bar{\imath}$ and C_A were obtained for each of the protein–dye mixtures. These are compared with the data of Klotz et al. (24) in Fig. 25. As seen therein, the values of $\bar{\imath}$ calculated from the sedimentation velocity experiments by this procedure are slightly and almost consistently higher than those obtained by equilibrium dialysis. Nevertheless, the agreement is really quite good. This is gratifying, since the procedures used are directly analogous to those employed by Chanutin and co-workers (35) and Velick and his co-workers (37–39) to obtain binding data by analysis of fractions withdrawn from tubes in the preparative centrifuge. Moreover, the small discrepancy between the sedimentation-velocity and equilibrium dialysis experiments can be reduced by about

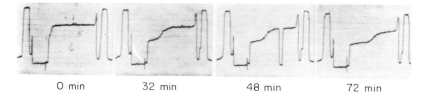

O min   32 min   48 min   72 min

FIG. 24. Sedimentation patterns of a mixture of bovine serum albumin and methyl orange recorded with a photoelectric scanner using light of wavelength 334 mμ. Constituent concentrations were 0.2 g (100 ml)$^{-1}$ of protein and 4.9×10^{-4} M methyl orange; 0.1 M phosphate buffer, pH 5.66, at 4°C; 59,780 rpm and indicated times after reaching speed. In the third pattern, as the photomultiplier slit assembly traversed the plateau containing both constituents, the nulling switch on the console was momentarily depressed so as to record the baseline and permit accurate readings of the total absorbance in the plateau. The baseline in that pattern coincides with that in the air space above the liquid columns showing zero optical density. [From Steinberg and Schachman (47).]

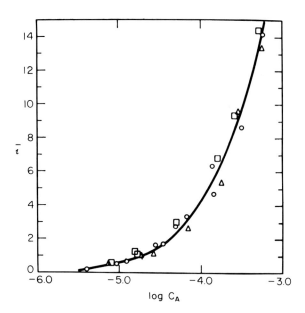

FIG. 25. Comparison of experimental results on the binding of methyl orange to bovine serum albumin. Plot of $\bar{\imath}$, the moles of methyl orange bound per mole of total protein, versus the logarithm of C_A, the equilibrium concentration of unbound methyl orange: (○) equilibrium dialysis data of Klotz et al. (24); (△) sedimentation equilibrium data of Steinberg and Schachman (47); and (□) sedimentation velocity experiments of Steinberg and Schachman (47). In all experiments, the buffer was 0.1 M phosphate, pH 5.66, and the temperature about 4°C. [From Steinberg and Schachman (47).]

50% if empirical correction is made for the slight depletion of the unbound methyl orange in the supernatant due to its sedimentation. The correction factor is obtained by sedimenting a control methyl orange solution devoid of protein at the same speed and for the same length of time as used for the dye–protein mixture, and measuring the decrease in dye concentration at the same position as used in the actual binding experiment.

Analysis of the residual discrepancy is a much more difficult problem. It is clear, however, that several chemical and hydrodynamic factors can cause adjustment of the concentration of unbound small molecule across the reaction boundary. These include readjustment of the chemical equilibria in the neighborhood of the boundary during differential transport of the complexes (*17*, *48*); possible existence of interacting flows (*49*), as reflected in cross-term diffusion coefficients; and the effects of radial dilution and pressure (*50*, *51*) on equilibrium position. A few words concerning the latter are in order. At the moderate to high rotor speeds used in sedimentation-velocity experiments, high pressures, of the order of 100–500 atm, may be generated at the cell bottom. If the molar-volume change upon reaction is sufficiently large, the value of the equilibrium constant will vary significantly along the centrifuge cell. In fact, for certain macromolecular interactions, this pressure effect may be enormous (*52*), with serious consequences for interpretation of sedimentation-velocity experiments (*50*, *51*). In the case of protein–small-molecule interactions, the molar-volume change is usually small, of the order of 10 ml mole^{-1} (*53*). Accordingly, for such systems, the pressure effect causes only minor interpretative difficulties. Large volume changes would undoubtedly preclude quantitative description of the chemical equilibria from sedimentation-velocity data. Having demonstrated the usefulness of the relatively simple and rapid method described above for obtaining binding data, Steinberg and Schachman proceeded to examine the feasibility of applying the concept of constituent sedimentation coefficient to this problem. In experiments on a series of albumin–methyl orange mixtures of known $\bar{c}_\mathscr{P}$ and $\bar{c}_\mathscr{A}$, both Schlieren and conventional absorption sedimentation-velocity patterns were photographed simultaneously. In a given experiment, the value of $\bar{s}_\mathscr{P}$ was determined from the rate of sedimentation of the reaction boundary in the Schlieren patterns, and $\bar{s}_\mathscr{A}$ was evaluated by applying the transport equation (143) to photodensitometer traces of the absorption patterns. But, since the protein preparation used in these experiments contained an appreciable amount of dimer, interpretation of the constituent mobilities is fraught with difficulties. Thus, $\bar{s}_\mathscr{A}$ provides a measure of the transport of all dye molecules, including dye bound to the dimeric form of the protein as well as that bound to the monomer. In contrast, $\bar{s}_\mathscr{P}$ measured from the maximum ordinate of the reaction boundary corresponds only to the monomeric species of the protein constituent. Furthermore, correction of $\bar{s}_\mathscr{P}$ so as to include the dimer is somewhat uncertain. Consequently, a strict comparison between the values of $\bar{s}_{\mathscr{A}}$ and those for $\bar{s}_{\mathscr{P}}$ is not justified, and their use to compute c_A from Eq. (137) or (138) is not warranted. It was possible, however, by application of the highly precise comparative technique used to obtain the data presented in Fig. 21, to demonstrate that the factor $(\bar{s}_{\mathscr{P}} - s_P)$ in Eq. (137) is negligibly small compared to other terms. This provides experimental justification for the use of Eq. (138) in evaluating the association equilibria. Despite the difficulties encountered in the interpretation of $\bar{s}_{\mathscr{P}}$ for serum albumin, these pioneering and carefully executed experiments provide the guidelines for future application of constituent sedimentation coefficients to the quantitative characterization of the interactions of other proteins with small molecules.

Sedimentation Equilibrium. In comparison with velocity sedimentation, equilibrium sendimentation of interacting systems is relatively uncomplicated. Thus, there is no disturbance of the chemical equilibria due to transport; interacting flows are of no consequence, and the pressure gradient through the cell is usually small at the relatively low centrifugal fields employed. As we saw in Chapter I, at sedimentation equilibrium, the distribution throughout the cell of each substance comprising the system is in accordance with a thermodynamic equation [Eq. (76) in the case of ideal solutes] which describes the concentration as a function of position in terms of the molecular weight and partial specific volume of the solute, the density of the solution, and the intensity of the centrifugal field. Moreover, chemical equilibrium between the reacting substances is attained at every position in the cell.

For a noninteracting mixture of macromolecule and small molecule at sedimentation equilibrium in a field of relatively low intensity, the small molecule will be practically uniformly distributed throughout the cell even though there is a large concentration gradient of the macromolecule. If, however, there is an interaction between the macromolecule and the small molecule, the *total* concentration of the latter will increase with distance from the center of rotation (i.e., with increasing concentration of macromolecule), since a larger amount will be bound at higher protein concentration. Despite this gradient in the *total* concentration of the small molecule, the concentration of unbound small molecule will be virtually constant throughout the cell.

These assertions are simply restatements, made within the context of macromolecule–small-molecule systems, of the conditions for sedimentation equilibrium expressed mathematically by Eqs. (68)–(71) and (74).

Since chemical equilibrium is attained at every position in the cell, the definition of $\bar{\imath}$ given by Eq. (120) can be applied at every position. Rearranging Eq. (120), asserting the virtual constancy of c_A, and introducing the functional notation, yields

$$\bar{c}_{\mathscr{A}}(r) = c_A + \bar{\imath}\bar{c}_{\mathscr{P}}(r) \tag{144}$$

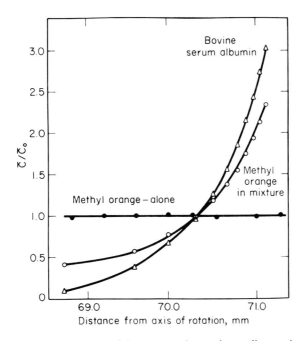

FIG. 26. Equilibrium distribution of the two constituents in a sedimentation-equilibrium experiment on a mixture of bovine serum albumin and methyl orange. Plot of relative concentration, $\bar{c}/\bar{c}_0$, vs. position, where $\bar{c}_0$ is the initial constituent concentration: (△) protein in the mixture; (○) methyl orange in the mixture; and (●) methyl orange in a control solution devoid of protein. Initial constituent concentrations: protein, 0.4 g/100 ml; methyl orange, 7.8×10^{-5} M. 0.1 M phosphate buffer, pH 5.66; 12,590 rpm; 18-hr sedimentation at 4°C. Distribution of methyl orange recorded with the photoelectric scanner at a wavelength of 440 mμ, the isosbestic point of methyl orange–serum albumin mixtures. Distribution of protein measured with interference optics. [From Steinberg and Schachman (47).]

in which $\bar{c}(r)$ is the constituent concentration at position r. According to this equation, a plot of $\bar{c}_{\mathscr{A}}(r)$ vs. $\bar{c}_{\mathscr{P}}(r)$ will give a straight line with an intercept on the abscissa equal to c_A and a slope equal to $\bar{\imath}$, provided the effect of pressure is indeed negligible and $\bar{\imath}$ is insensitive to total protein concentration (see footnote 4, this chapter).

In order to apply Eq. (144), it is necessary to determine the equilibrium distributions of both constituents in the same experiment. In their equilibrium experiments on the binding of methyl orange by serum albumin, Steinberg and Schachman used the photoelectric scanner to measure the distribution of *total* methyl orange and interference optics to measure the protein. The distributions obtained in a typical experiment are displayed in Fig. 26. It is im-

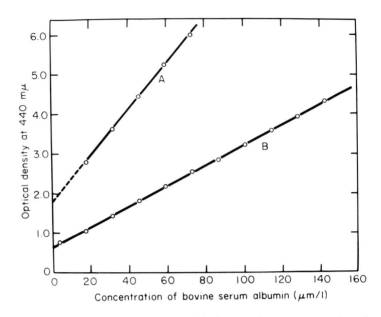

FIG. 27. Linear relationship between $\bar{c}_\mathscr{A}(r)$, the constituent concentration of methyl orange (expressed as optical density at 440 mμ) and $\bar{c}_\mathscr{P}(r)$, the constituent concentration of bovine serum albumin at sedimentation equilibrium: (A) initial constituent concentrations of protein and methyl orange were 46.5 and 188.8 μmole liter^{-1}, respectively; (B) 46.5 and 75.3. Values of $\bar{\imath}$ and c_A computed from intercept and slope: (A) 2.56 and 69.7 μmole liter^{-1}, respectively; (B) 1.07 and 25.6. Other conditions as in Fig. 26. [From Steinberg and Schachman (47).]

mediately apparent that only part of the methyl orange was bound to protein; otherwise, the distributions of the two constituents would be identical. For comparison, the equilibrium distribution of methyl orange obtained in the control experiment on a solution containing the same constituent concentration of dye but no protein is also displayed in Fig. 26. As expected, the equilibrium concentration of methyl orange was virtually constant throughout the cell in the control experiment. Thus, any change in the concentration of the dye in its protein mixture is attributable directly to the interaction between the macromolecules and the small dye ions. Finally, we note that the "hinge points" for the two constituents in the protein–dye mixture coincide, i.e., the position in the cell at which the constituent concentration at sedimentation equilibrium equals that in the initial mixture, is the same for both protein and dye. This must be so if the condition for chemical equilibrium is to be satisfied.

A further test of the theory is provided by the linear relationship between $\bar{c}_\mathscr{A}(r)$ and $\bar{c}_\mathscr{P}(r)$ illustrated in Fig. 27. Values of $\bar{\imath}$ and C_A obtained from the slopes and intercepts of such plots [Eq. (144)] are compared with the equilibrium dialysis measurements of Klotz and his co-workers (24) in Fig. 25. The agreement is seen to be excellent.

From both theoretical and experimental points of view, sedimentation methods have much to commend them for the evaluation of interactions between large and small molecules. The theory of sedimentation equilibrium of interacting systems is rigorous and the application of the method straightforward. Although the theory of sedimentation velocity is not as well understood, experimentally, it is more rapid and often simpler to apply. Taken together, the two approaches constitute a powerful means for analysis of many interacting systems of biological interest. Steinberg and Schachman have given a critical discussion of the advantages and limitations of the two methods and the prospects of extending them to the study of interactions between different types of macromolecules.

INTERACTIONS BETWEEN DIFFERENT MACROMOLECULES

Although the weak-electrolyte moving-boundary theory was originally formulated for macromolecule–small-ion interactions, its application is not restricted to such systems. The theory can also be used for the quantitative interpretation of the electrophoretic behavior of certain systems in which different macromolecules interact with each other. Specifically, it is applicable to simple systems of the type $nA + mB \rightleftharpoons A_nB_m$, in which only a *single* complex is formed in significant concentration. When two or more complexes are formed, rigorous analysis in terms of equilibrium constants becomes impossible. This is so because there are more unknowns than equations relating constituent quantities to the mobilities and concentrations of the various species in equilibrium (21). The problem of analyzing multiple equilibria was tractable in the case of macromolecule–small-ion interactions because the reasonable assumption could be made that binding of small ions changes the net charge, but not the frictional coefficient, of the macromolecule. Unfortunately, this assumption is not generally applicable to interactions between macromolecules. On the other hand, it should be noted that, even though rigorous analysis hinges upon formation of a single complex, it is not necessary that the stoichiometry of the complex be 1:1 on a molar basis (7). Finally, interactions for which the electrophoretic mobility of the complex is intermediate in value between those of the two macromolecular reactants are particularly amenable to analysis (7, 18–22). This is illustrated below by the electrophoretic demonstration of the specific Michaelis–Menten complex between pepsin and bovine serum albumin.

Consider the reversible enzyme–protein substrate interaction $E + S \rightleftharpoons ES$, in which the enzyme E has a greater mobility than the substrate S, and the enzyme–substrate complex ES is of intermediate mobility. The moving-boundary system indicated for characterization of such an interaction is presented in Fig. 28. The initial boundaries are formed between buffer and

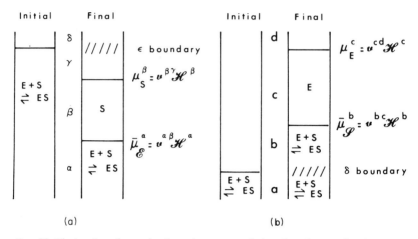

Fig. 28. Electrophoretic moving-boundary system designed to characterize the reversible interaction between pepsin, E, and serum albumin, S, to form the specific-substrate complex ES: (a) descending limb; (b) ascending limb. This system is representative of interactions for which the electrophoretic mobility of the *single* complex is intermediate in value between those of the macromolecular reactants. See text for details. The ions of the buffer exist throughout the moving-boundary system and are not indicated.

buffered solution containing known constituent concentrations of $\mathscr{E}$ and $\mathscr{S}$. For sufficiently large rates of establishment of equilibrium, the electrophoretic process will generate a system of two moving boundaries in both the descending and ascending limbs of the Tiselius cell. In the descending limb, the more rapidly migrating $\mathscr{E}$ constituent moves downwards in the initial equilibrium mixture (α phase), leaving behind uncombined S in the β phase. Thus, we see that the leading $\alpha\beta$ boundary is a reaction boundary in which the equilibrium is adjusted as rapidly as required by electrophoretic separation of E and ES; the rate of migration of its first moment gives the constituent mobility of the enzyme in the α phase, which will be smaller than the mobility of uncombined enzyme. In contrast, the slower-migrating $\beta\gamma$ boundary corresponds to uncombined substrate. In the ascending limb, uncombined E migrates out of the adjusted equilibrium mixture of the b phase into fresh buffer, thereby giving rise to the leading cd boundary. The slower moving bc boundary is a reaction boundary whose mean rate of migration gives the constituent mobility of substrate in the b phase. If the interaction is sufficiently strong, this boundary will be bimodal (48). Finally, it is important to note that the area under the leading ascending boundary will be less than that of the leading descending one. This is so because the former corresponds to uncombined enzyme and the latter to total enzyme plus combined substrate.

The electrophoretic patterns can be interpreted quantitatively in terms of the dissociation constant K of the enzyme–substrate complex:

$$K = \frac{c_E^\alpha c_S^\alpha}{c_{ES}^\alpha} = \frac{c_E^\alpha(\bar{c}_\mathscr{S}^\alpha - \bar{c}_\mathscr{E}^\alpha + c_E^\alpha)}{\bar{c}_\mathscr{E}^\alpha - c_E^\alpha} \tag{145}$$

where the only unknown quantity, c_E^α, can be derived from mobility and area measurements as follows: The constituent quantities are defined as

$$\bar{c}_\mathscr{E}^\alpha = c_E^\alpha + c_{ES}^\alpha \tag{146}$$

$$\bar{c}_\mathscr{S}^\alpha = c_S^\alpha + c_{ES}^\alpha \tag{147}$$

$$\bar{\mu}_\mathscr{E}^\alpha = (c_E^\alpha/\bar{c}_\mathscr{E}^\alpha)\mu_E^\alpha + (c_{ES}^\alpha/\bar{c}_\mathscr{E}^\alpha)\mu_{ES}^\alpha \tag{148}$$

$$\bar{\mu}_\mathscr{S}^\alpha = (c_S^\alpha/\bar{c}_\mathscr{S}^\alpha)\mu_S^\alpha + (c_{ES}^\alpha/\bar{c}_\mathscr{S}^\alpha)\mu_{ES}^\alpha \tag{149}$$

Eliminating c_{ES}^α between Eqs. (146) and (147) and $c_{ES}^\alpha \mu_{ES}^\alpha$ between Eqs. (148) and (149) yields

$$\bar{c}_\mathscr{E}^\alpha - c_E^\alpha = \bar{c}_\mathscr{S}^\alpha - c_S^\alpha \tag{150}$$

and

$$\bar{\mu}_\mathscr{E}^\alpha \bar{c}_\mathscr{E}^\alpha - c_E^\alpha \mu_E^\alpha = \bar{\mu}_\mathscr{S}^\alpha \bar{c}_\mathscr{S}^\alpha - c_S^\alpha \mu_S^\alpha \tag{151}$$

Now, the moving-boundary equation (107) for $\mathscr{S}$ across the $\alpha\beta$ boundary is

$$\frac{\bar{\mu}_\mathscr{S}^\alpha \bar{c}_\mathscr{S}^\alpha}{\mathscr{H}^\alpha} - \frac{\mu_S^\beta c_S^\beta}{\mathscr{H}^\beta} = v^{\alpha\beta}(\bar{c}_\mathscr{S}^\alpha - c_S^\beta) \tag{152}$$

Assuming $\mathscr{H}^\alpha = \mathscr{H}^\beta$ and noting that $v^{\alpha\beta}\mathscr{H}^\alpha = \bar{\mu}_\mathscr{E}^\alpha$,

$$\bar{\mu}_\mathscr{S}^\alpha \bar{c}_\mathscr{S}^\alpha - \mu_S^\beta c_S^\beta = \bar{\mu}_\mathscr{E}^\alpha(\bar{c}_\mathscr{S}^\alpha - c_S^\beta) \tag{153}$$

By eliminating $\bar{\mu}_\mathscr{S}^\alpha \bar{c}_\mathscr{S}^\alpha$ and c_S^α between Eqs. (150), (151), and (153) and assuming $\mu_S^\beta = \mu_S^\alpha$, we obtain the following equation which relates the equilibrium concentration of uncombined enzyme to overall composition and electrophoretically measurable quantities:

$$c_E^\alpha = (\bar{c}_\mathscr{E}^\alpha + c_S^\beta - \bar{c}_\mathscr{S}^\alpha)(\bar{\mu}_\mathscr{E}^\alpha - \mu_S^\alpha)/(\mu_E^\alpha - \mu_S^\alpha) \tag{154}$$

The quantities μ_E^α and μ_S^α are taken as the mobilities of the two constituents in their pure solutions; $\bar{\mu}_\mathscr{E}^\alpha$ is the mean mobility of the descending $\alpha\beta$ boundary; $\bar{c}_\mathscr{E}^\alpha$ and $\bar{c}_\mathscr{S}^\alpha$ are fixed by the overall composition of the equilibrium mixture used to form the initial boundary; and c_S^β is determined from the area of the descending $\beta\gamma$ boundary. Having thus obtained the equilibrium

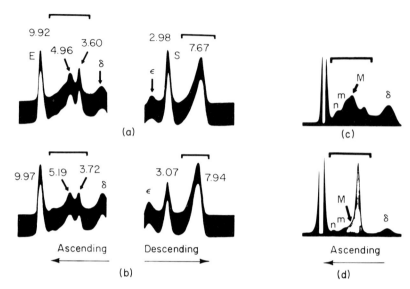

Fig. 29. Electrophoretic patterns of mixtures of pepsin and serum albumin, ionic strength 0.1 phosphate buffer, pH 5.35: (a) 1% pepsin + 1% albumin, electric field applied 25 min after mixing at 0°C, time of electrophoresis, 90 min; (b) 1% pepsin + 1% albumin after 5 hr of dialysis against buffer at 0°C, (c) 1.22% pepsin + 0.595% albumin; (d) 1.14% pepsin + 0.560% albumin + 3 × 10^{-3} M carbobenzoxy-α-L-glutamyl-L-tyrosine. Reaction boundaries are indicated by brackets, and the boundaries corresponding to uncombined serum albumin, S, and pepsin, E, are indicated in (a). In (a) and (b), the mobilities, $(-10^5) \times$ cm^2 sec^{-1} V^{-1}, and apparent mobilities are shown above or beside the corresponding peaks. The descending mobilities of pepsin and albumin in solutions of pure constituents are $-9.42 \pm 0.05 \times 10^{-5}$ cm^2 sec^{-1} V^{-1} and $-2.85 \pm 0.03 \times 10^{-5}$, respectively; the apparent ascending mobilities are $-9.68 \pm 0.02 \times 10^{-5}$ and $-2.99 \pm 0.02 \times 10^{-5}$. [Adapted from Cann and Klapper (18).]

concentration of uncombined enzyme, the dissociation constant of the enzyme–substrate complex is computed using Eq. (145).

In the pepsin–serum albumin experiments (18), mixtures of enzyme with substrate were examined electrophoretically under conditions such that proteolysis had negligible effect on electrophoretic patterns obtained shortly after mixing.[6] Several representative patterns are displayed in Fig. 29. It is apparent from Fig. 29a that the patterns possess all of the features described

---

[6] In preliminary experiments, the rate of proteolysis was followed by dialyzing the reaction mixture for varying times prior to electrophoretic analysis. Comparison of Figs. 29a and 29b illustrate the slow decrease in area of the descending albumin boundary. These and other observations permitted the conclusion that proteolysis has a negligible effect on patterns obtained immediately after mixing the enzyme and substrate.

above for reversible formation of a complex of intermediate mobility. Supporting evidence for a pepsin–albumin complex of intermediate mobility was provided by experiments in which the ratio of constituent concentrations of the two proteins was varied at constant total protein concentration. The constituent mobilities changed in accordance with expectation, and the mobilities of the slow descending and rapid ascending boundaries were only slightly altered.

The bimodality of the ascending reaction boundary in patterns obtained at low and moderate initial mixing ratios of pepsin to albumin indicates that a single type of pepsin–albumin complex predominates under these circumstances. The simplifying assumption can be made that the faster-migrating peak in the reaction boundary (the peak with apparent mobility 4.96 in Fig. 29a and designated as M in Figs. 29c and 29d) is a qualitative representation of the complex. Although the areas and migration velocities of the peaks in a reaction boundary cannot be identified in the conventional manner with the concentrations and mobilities of the components in the solution (48), the relative area of the M peak proved to be an empirical index of the extent of binding of albumin by pepsin. Increasing the ratio of pepsin to albumin results in the appearance of two additional peaks in the reaction boundary, the m and n peaks of Figs. 29c and 29d, which migrate more rapidly than the M peak and which predominate over the latter at the highest ratio examined. The presence of several maxima in the reaction boundary at the higher ratios indicates the existence of several different types of complexes under these conditions.

The various patterns, except for those obtained at the highest ratio, have been interpreted quantitatively in terms of the equilibrium constant for dissociation of the enzyme–substrate complex. The computations of K with the aid of Eqs. (145) and (154), which are summarized in part A of Table II, assume that the complex contains one molecule of pepsin and one of albumin. This assumption is justified by the result that the same value of K was obtained for initial pepsin–albumin ratios $\bar{c}_{\mathscr{E}}^{\alpha}/\bar{c}_{\mathscr{S}}^{\alpha}$ of 0.39 and 0.73, conditions under which the m and n peaks are negligibly small. Furthermore, the concentrations of combined pepsin and combined albumin are in fairly good agreement. The situation is somewhat different at a pepsin–albumin ratio of 2.1–2.2. The value of K is about 60% smaller than at the lower ratios, and the concentration of combined pepsin is considerably greater than the concentration of combined albumin, which correlates with the appearance of the additional m and n peaks in the ascending reaction boundary. These results lead us to conclude that the complex formed at low and moderate pepsin–albumin ratios contains a single molecule each of pepsin and albumin and that the additional complexes formed at higher ratios are apparently higher-order ones containing a greater amount of pepsin than albumin. It should be borne

TABLE II

Equilibrium Constants for Dissociation of Pepsin–Serum Albumin Complex at pH 5.35 and 1°

	$\bar{C}_\varepsilon^\alpha$	$\bar{C}_\gamma^\alpha$	C_S^β	C_E^α	$\bar{\mu}_\varepsilon^\alpha$ (-10^5 × cm^2 sec^{-1} V^{-1})	K (10^5 × mole liter^{-1})
$\bar{C}_\varepsilon^\alpha/\bar{C}_\gamma^\alpha$		(10^4 × mole liter^{-1})				
A. Ionic strength 0.1: 0.39	0.849	2.19	1.80	0.246	6.38	6.49
$\mu_E^\alpha = -9.42 \times 10^{-5}$, 0.73	1.42	1.94	1.39	0.540	6.93	6.51
$\mu_S^\alpha = -2.85 \times 10^{-5}$ 2.1	2.54	1.24	0.785	1.48	7.51	2.52
2.2	2.79	1.27	0.806	1.68	7.00	2.42
B. Ionic strength 0.2: 2.2	2.83	1.27	1.17	2.11	6.82	16.1
$\mu_E^\alpha = -7.88 \times 10^{-5}$,						
$\mu_S^\alpha = -3.19 \times 10^{-5}$						

[a] Pentex bovine serum albumin lot 1207. For Pentex lot 9F07, the Michaelis–Menten constant K_m was equal to 1.3×10^{-5} mole liter^{-1} at ionic strength 0.1 and pH 5.35 (literature value, 1.5×10^{-5} at pH 1.5) and 7.4×10^{-5} at ionic strength 0.2.

in mind, however, that while it is important to recognize the probable existence of a variety of complexes at the higher ratios, the theory is rigorously applicable only at the low and moderate ratios where a single complex is formed. It is satisfying that the dissociation constant obtained for the latter ratios is of the same order of magnitude as the Michaelis–Menten constant. This is to be expected, since the rate of formation of hydrolytic products k_3 is small under the conditions of the electrophoretic experiments.

Several independent lines of evidence indicate that the one-to-one complex is the Michaelis–Menten complex capable of being activated to give rise to reaction products and that the higher-order complexes are also specific: (1) parallelism between the effect of ionic strength on K_m and on complex formation (see Table II); (2) inhibition of complex formation by the peptic digest of bovine serum albumin; (3) inhibition by the low-molecular-weight substrate carbobenzoxy-α-L-glutamyl-L-tyrosine, which competes with albumin for the active site on the enzyme molecule (compare Figs. 29c and 29d); and (4) lack of detectable complex formation between pepsin and ovalbumin under the same conditions as the pepsin–serum albumin experiments.

The ability to interpret the electrophoretic patterns of this interacting system in a quantitative fashion has made possible investigations into the effect of acetyl-L-tryptophan and fatty acids on specific complexing and the rate of proteolysis (*19, 20*). The results bear upon the mechanism of proteolysis and the structure of albumin. Finally, these experiments constitute the first unambiguous physical demonstration of a Michaelis–Menten complex between a proteolytic enzyme and its macromolecular substrate. Previously, the existence of the Michaelis–Menten complex had been shown only for low-molecular-weight substances (*54–58*).

REFERENCES

1. H. Svensson, *Acta Chem. Scand.* **2**, 841 (1948).
2. R. A. Alberty and J. C. Nichol, *J. Am. Chem. Soc.* **70**, 2297 (1948).
3. R. A. Alberty, *J. Am. Chem. Soc.* **72**, 2361 (1950).
4. E. B. Dismukes and R. A. Alberty, *J. Am. Chem. Soc.* **76**, 19 (1954).
5. J. C. Nichol, E. B. Dismukes and R. A. Alberty, *J. Am. Chem. Soc.* **80**, 2610 (1958).
5a. A. Tiselius, *Nova Acta Regiae Soc. Sci. Upsaliensis* [4] **7**, No. 4 (1930).
6. J. G. Kirkwood, *in* "Proteins, Amino Acids and Peptides as Ions and Dipolar Ions" (E. J. Cohn and J. T. Edsall, eds.), Chapter 12. Reinhold, New York, 1943.
7. L. G. Longsworth, *in* "Electrophoresis Theory, Methods and Applications" (M. Bier, ed.), Chapter 3. Academic Press, New York, 1959.
8. V. P. Dole, *J. Am. Chem. Soc.* **67**, 1119 (1945).
9. H. Svensson, *J. Am. Chem. Soc.* **72**, 1974 (1950).
10. J. R. Cann, *in* "Treatise on Analytical Chemistry" (I. M. Kolthoff and P. J. Elving, eds.), Vol. 2, Part 1, Chapter 28. Wiley (Interscience), New York, 1961.
11. L. G. Longsworth, *J. Am. Chem. Soc.* **65**, 1755 (1943).
12. R. F. Smith and D. R. Briggs, *J. Phys. Colloid Chem.* **54**, 33 (1950).
13. R. A. Alberty and H. H. Marvin, Jr., *J. Phys. Colloid Chem.* **54**, 47 (1950).
14. I. Z. Steinberg and H. K. Schachman, *Biochemistry* **5**, 3728 (1966).
15. R. A. Alberty and H. H. Marvin, Jr., *J. Am. Chem. Soc.* **73**, 3220 (1951).
16. K. Schilling, *Acta Chem. Scand.* **11**, 1103 (1957).
17. J. R. Cann and W. B. Goad, *J. Biol. Chem.* **240**, 148 (1965).
18. J. R. Cann and J. A. Klapper, Jr., *J. Biol. Chem.* **236**, 2446 (1961).
19. J. R. Cann, *J. Biol. Chem.* **237**, 707 (1962).
20. J. A. Klapper, Jr. and J. R. Cann, *Arch. Biochem. Biophys.* **108**, 531 (1964).
21. L. Ehrenpreis and R. C. Warner, *Arch. Biochem. Biophys.* **61**, 38 (1956).
22. S. N. Timasheff and J. G. Kirkwood, *J. Am. Chem. Soc.* **75**, 3124 (1953).
23. L. G. Longsworth and C. F. Jacobsen, *J. Phys. Colloid Chem.* **53**, 126 (1949).
24. I. M. Klotz, F. M. Walker and R. B. Pivan, *J. Am. Chem. Soc.* **68**, 1486 (1946).
25. G. Scatchard, J. S. Coleman, and A. L. Shen, *J. Am. Chem. Soc.* **79**, 12 (1957).
26. F. Karush, *J. Am. Chem. Soc.* **78**, 5519 (1956).
27. A. Nisonoff and D. Pressman, *J. Immunol.* **80**, 417 (1958).
28. J. D. Teresi and J. M. Luck, *J. Biol. Chem.* **194**, 823 (1952).
28a. J.-P. Changeux, J. C. Gerhart, and H. K. Schachman, *Biochemistry* **7**, 531 (1968).
28b J. C. Gerhart and H. K. Schachman, *Biochemistry* **7**, 538 (1968).
28c. J.-P. Changeux and M. M. Rubin, *Biochemistry* **7**, 553 (1968).
29. J. Steinhardt and S. Beychok, *in* "The Proteins Composition, Structure and Function" (H. Neurath, ed.), 2nd ed. Vol. II, Chapter 8. Academic Press, New York, 1964.
30. J. R. Cann and R. A. Phelps, *J. Am. Chem. Soc.* **77**, 4266 (1955).
31. R. A. Phelps and J. R. Cann, *Biochim. Biophys. Acta* **23**, 149 (1957).

32. S. Nakamura, K. Takeo, I. Sasaki and M. Murata, *Nature* **184**, 638 (1959).
33. E. I. McDougall and R. L. M. Synge, *Brit. Med. Bull.* **22**, 115 (1966).
34. M. H. Bickel and D. Bovet, *J. Chromatog.* **8**, 466 (1962).
35. A. Chanutin, S. Ludewig and A. V. Masket, *J. Biol. Chem.* **143**, 737 (1942).
36. I. M. Klotz, *Arch. Biochem.* **9**, 109 (1946).
37. S. F. Velick, J. E. Hayes, Jr., and J. Harting, *J. Biol. Chem.* **203**, 527 (1953).
38. J. E. Hayes, Jr. and S. F. Velick, *J. Biol. Chem.* **207**, 225 (1954).
39. S. F. Velick, *in* "The Mechanism of Enzyme Action, Symposium" (W. D. McElroy and B. Glass, eds.), p. 491. Johns Hopkins Press, Baltimore, Maryland, 1954.
40. W. Gilbert and B. Müller-Hill, *Proc. Natl. Acad. Sci. U.S.* **56**, 1891 (1966).
41. W. Gilbert and B. Müller-Hill, *Proc. Natl. Acad. Sci. U.S.* **58**, 2415 (1967).
42. E. Fredericq and H. Neurath, *J. Am. Chem. Soc.* **72**, 2684 (1950).
43. W. L. Hughes, Jr., *J. Am. Chem. Soc.* **69**, 1836 (1947).
44. W. L. Hughes, Jr., *Symp. Quant. Biol. Cold Springs Harbor* **14**, 79 (1950).
45. J. Monod, J.-P. Changeux and F. Jacob, *J. Mol. Biol.* **6**, 306 (1963).
46. J. Monod, J. Wyman and J.-P. Changeux, *J. Mol. Biol.* **12**, 88 (1965).
47. I. Z. Steinberg and H. K. Schachman, *Biochemistry* **5**, 3728 (1966).
48. G. A. Gilbert and R. C. Ll. Jenkins, *Proc. Roy. Soc.* (*London*) **A253**, 420 (1959).
49. L. J. Gosting, *Advan. Protein Chem.* **11**, 429 (1956).
50. G. Kegeles, L. Rhodes and J. L. Bethune, *Proc. Natl. Acad. Sci. U.S.* **58**, 45 (1967).
51. L. F. Ten Eyck and W. Kauzmann, *Proc. Natl. Acad. Sci. U.S.* **58**, 888 (1967).
52. R. Josephs and W. F. Harrington, *Proc. Natl. Acad. Sci. U.S.* **58**, 1587 (1967).
53. R. M. Rosenberg and I. M. Klotz, *J. Am. Chem. Soc.* **77**, 2590 (1955).
54. K. Keilin and T. Mann, *Proc. Roy. Soc.* (*London*) **B122**, 119 (1937).
55. K. G. Stern, *J. Biol. Chem.* **114**, 473 (1939).
56. B. Chance, *J. Biol. Chem.* **151**, 553 (1943).
57. B. Chance, *Advan. Enzymol.* **12**, 153 (1951).
58. D. G. Doherty and F. Vaslow, *J. Am. Chem. Soc.* **74**, 931 (1952).

III/ ANALYTICAL SOLUTION OF APPROXIMATE CONSERVATION EQUATIONS

ANALYTICAL SOLUTION OF APPROXIMATE CONSERVATION EQUATIONS

The foregoing formulation of the weak-electrolyte moving-boundary theory made use of an integral form of the conservation equation to compute the rate of migration of a reaction boundary and the net changes in concentration of constituents across the boundary, without making any statements as to what is occurring within the boundary itself. As we have seen, the concepts thus elaborated permit quantitative interpretation of the electrophoretic patterns shown by certain types of interacting systems in terms of equilibrium constants and other parameters. On the other hand, the theory does not describe the effect of the interaction on the shape of the reaction boundary. Consider, for example, the electrophoretic patterns shown by a mixture of pepsin and serum albumin (Fig. 29a). The weak-electrolyte moving-boundary theory predicts that the rate of migration of the first moment of the descending or the ascending reaction boundary is the constituent velocity of pepsin in the α phase or albumin in the b phase, respectively. However, by its very nature, the theory gives us no reason to expect the descending reaction boundary to be unimodal and the ascending one bimodal. Such insight has come only from solution of the conservation equations for reacting systems of the type $A + B \rightleftharpoons C$, in which equilibrium is established instantaneously. In fact, prior to the theoretical work of Gilbert (*1, 2*) and Gilbert and Jenkins (*3, 4*) on the mass transport of reacting systems, an intuitive approach to these problems was necessary; and it was widely, but erroneously, held that rapidly equilibrating systems would necessarily show unimodal reaction boundaries. Bimodal boundaries were usually interpreted in terms of kinetically controlled processes. While there are, of course, many authenticated examples of slowly attained macromolecular equilibria, it is becoming increasingly clear that a variety of rapidly equilibrating reactions can and do give rise to multiple electrophoretic and ultracentrifugal peaks and zones. The implications of these new insights for conventional analytical applications of electrophoresis, ultracentrifugation, and chromatography are elaborated in Chapter VI. Here, we are concerned with the application of these methods to the quantitative characterization of macromolecular association–dissociation reactions, interactions between different macromolecules, and allosteric interactions. The central role of such interactions in current biological thought warrants a detailed mathematical analysis of at least one of these systems. We choose to do this for the macromolecular association–dissociation reaction because Gilbert's treatment of this problem represents a major advance in understanding the mass transport of reacting systems. Moreover, essentially the same mathematical concepts are employed in the theoretical elaboration of the somewhat more complex situation presented by interactions of different macromolecules with each other.

Analytical solution of the exact differential equations describing the mass transport of reversibly reacting systems is an extremely difficult mathematical problem. This is so even when the equations are linear, as in the case of macromolecular isomerization, $A \underset{k_2}{\overset{k_1}{\rightleftharpoons}} B$, with rates of interconversion comparable to the rate of electrophoretic separation of the isomers. (See pp. 48–50, Chapter II, for formulation of theory.) During electrophoresis of this system, the changes in concentrations of A and B with time of electrophoresis and position in the electrophoresis column are described by the set of transport-interconversion equations (82a) and (82b). Analytical expressions for the Fourier transforms of the concentrations were obtained from these equations by Cann et al. (5). Although exact evaluation of the Fourier integrals which give the concentrations was not possible, an approximate solution yielded the prediction that, for certain conditions, resolution of the Schlieren patterns into two peaks will occur for times of electrophoresis of the order of the half-time of reaction. The mathematical difficulties evidently revolve around the second-order terms which describe the effect of diffusion in the differential equations. If it is assumed that diffusion is negligible, Eqs. (82a) and (82b) reduce to the first-order equations

$$\frac{\partial c_1}{\partial t} = -\mu_1 E \frac{\partial c_1}{\partial x} - k_1 c_1 + k_2 c_2 \qquad (155a)$$

$$\frac{\partial c_2}{\partial t} = -\mu_2 E \frac{\partial c_2}{\partial x} + k_1 c_1 - k_2 c_2 \qquad (155b)$$

Van Holde (6) has obtained an analytical solution of these approximate transport-interconversion equations which predicts that under some circumstances the electrophoretic patterns may show three peaks. This result is in agreement with the more general numerical solution of the exact differential equations to be described in the next chapter.

The rationale for the assumption that diffusion can be ignored in treating the mass transport of interacting systems is that, whereas the effect of an external force on the shape of a moving boundary is directly proportional to time, the effect of diffusion varies with the square root of time. It seems reasonable, therefore, to isolate the effect of transport due to the electric or centrifugal field from the effect of diffusion, and to calculate the asymptotic form of the boundary. The latter should reveal the main influence of the interaction on boundary shape, even though it may not allow exact comparison with experiment.

The assumption that diffusion can be ignored has of necessity been made by Gilbert (*1, 2*) in his theory of reversible polymerization of a single macromolecule and by Gilbert and Jenkins (*4*) in their treatment of interactions between different macromolecules. However, even with this simplification, another difficulty is encountered, in that the differential equations are nonlinear. Fortunately, however, similarity solutions of these equations can be obtained. Since the similarity transformation is a powerful procedure for solving certain types of nonlinear partial differential equations, a brief statement of the method follows. The reader is referred to Ames (*7*) for a thorough treatment of the general theory and pertinent literature references.

THE SIMILARITY TRANSFORMATION

The physical meaning of the term "similarity" relates to internal or self-similitude of a problem. Thus, for electrophoresis of an interacting system, "similar" solutions of the conservation equations are those for which the electrophoretic patterns computed at two different times of electrophoresis differ only by a scale factor. As we shall see, such solutions apply only to the case of an infinitely sharp initial boundary. The mathematical interpretation of the term "similarity" is a transformation of variables, so carried out, that a reduction in the number of independent variables is achieved. In addition, the dependent variable may be changed into some new function. Consider an electrophoretic process described by a single partial differential equation in the dependent variable $c(x, t)$ and the two independent variables x and t, along with certain initial and boundary conditions. We attempt to consolidate the two independent variables appropriately into a single independent variable $\eta(x, t)$ such that the partial differential equation is reduced to an ordinary differential equation in $c(\eta)$ and which does not involve x and/or t separately. Such similarity transformations can be used only when (1) the original initial and boundary conditions can be consolidated to give the number of boundary conditions required for a unique solution of the transformed equation, and (2) these latter conditions are expressible in terms of the similarity variable η alone. The solution of the transformed equation is called a similarity solution.

In general, one may search for different classes of similarity transformations, such as, for example, $\eta = f(x^n/t^m)$ and $\eta = f(x + \lambda t)$. Transformations of the first class yield solutions which conserve mass during transport, but are usually applicable to only a single set of initial conditions; those of the second class yield solutions which are wave functions applicable to any initial conditions. Depending upon the nature of the problem, no similarity variables, one similarity variable, or many similarity variables may exist for a particular class of transformation. For example, it can be shown that either of the two similarity variables $\eta = x^2/t$ or $\eta = x/t^{1/2}$ may be used to solve Fick's second law of diffusion, irrespective of whether the diffusion coefficient is a constant (7, 8). Reduction of Fick's second law to an ordinary differential equation in the second of these similarity variables is called the *Boltzmann transformation*, and its use will be illustrated shortly. In any case, it must be borne in mind that a given similarity solution usually applies to only a single set of initial and boundary conditions, and one must determine what those conditions are.

There are several methods available to aid in the search for appropriate similarity variables, but we shall confine ourselves to a statement of a method which has proved particularly useful in transport problems and some areas of fluid mechanics—namely, the method of one-parameter groups of transformations. This method is based upon the invariance under a group of transformations of a system of partial differential equations. The meaning of "invariance under a group of transformations" is clarified by the following illustration: Consider Fick's second law of diffusion for the case in which the diffusion coefficient is independent of solute concentration:

$$(\partial c/\partial t)_x = D(\partial^2 c/\partial x^2)_t \tag{156}$$

If we make the transformations $\bar{x} = ax$ and $\bar{t} = a^2 t$, the partial differential equation becomes

$$(\partial c/\partial \bar{t})_{\bar{x}} = D(\partial^2 c/\partial \bar{x}^2)_{\bar{t}} \tag{157}$$

which has the same form as before. Thus, Fick's second law is invariant under this particular group of transformations, whose single parameter is a. Now, we search for a solution which is also invariant under the same group of transformations. This will be a similarity solution; and the similarity variables are found by the following general procedure.

Let Σ be a system of partial differential equations given by $\Phi_j = 0$ ($j = 1, 2, \ldots, n$) in which x_i ($i = 1, 2, \ldots, m$) and y_j ($j = 1, 2, \ldots, n$) are the independent and dependent variables, respectively. Let a group Γ_1 consisting of a set of transformations be defined as

$$\bar{x}_1 = a^{\alpha_1}x_1, \qquad \bar{x}_r = a^{\alpha_r}x_r, \ldots \qquad (r = 2, 3, \ldots, m)$$
$$\bar{y}_j = a^{\gamma_j}y_j \qquad (j = 1, 2, \ldots, n) \tag{158}$$

where the parameter $a \neq 0$ is real and α_r and γ_j are to be determined from the conditions that the system Σ be invariant under Γ_1. [Rigorously: constant conformally (absolutely) invariant; see footnote on p. 136 of Ames (7)]. This requirement gives rise to a set of simultaneous algebraic equations in α_r and γ_j, whose nontrivial solutions generate similarity variables which are invariants of Γ_1. Suppose x_1 is the independent variable to be eliminated. Solution of the algebraic equations gives rise to two cases: If $\alpha_1 \neq 0$, the invariants of Γ_1 are the similarity variables

$$\eta_r = x_r/x_1^{\beta_r}, \qquad \beta_r = \alpha_r/\alpha_1 \qquad (r = 2, 3, \ldots, m)$$
$$f_j(\eta_2, \eta_3, \ldots, \eta_m) = y_j(x_1, x_2, \ldots, x_m)/x_1^{\gamma_j/\alpha_1} \qquad (j = 1, 2, \ldots, n) \tag{159}$$

If $\alpha_1 = 0$ and the algebraic equations have a nontrivial solution, then we may choose a new group Γ_2 consisting of

$$\bar{x}_1 = x_1 + \ln a, \qquad \bar{x}_r = a^{\alpha_r}x_r, \qquad \bar{y}_j = a^{\gamma_j}y_j \tag{160}$$

and the invariants of the group are

$$\eta_r = \frac{x_r}{\exp(\alpha_r x_1)} \qquad (r = 2, 3, \ldots, m)$$
$$f_j(\eta_2, \ldots, \eta_m) = \frac{y_j(x_1, \ldots, x_m)}{\exp[\gamma_j x_1]} \qquad (j = 1, 2, \ldots, n) \tag{161}$$

This general procedure often permits (1) reduction of a partial differential equation in two independent variables to an ordinary differential equation in the similarity variable η; (2) reduction of a set of partial differential equations in two independent variables to a set of ordinary differential equations in η; and (3) by repeated application, reduction of a partial differential equation in more than two independent variables first to fewer independent variables and, finally, to an ordinary differential equation.

Before proceeding to the application of these concepts to solution of the approximate, nonlinear partial differential equations describing the electrophoresis and sedimentation of certain types of interacting systems, their utility and limitations will be illustrated by the solution of two linear transport problems. These are (1) Fick's second law of diffusion and (2) ideal electrophoresis or sedimentation to the rectilinear and uniform-field approximations of an inherently homogeneous, noninteracting macromolecule with the assumption that diffusion is negligible. The latter treatment will provide a theoretical framework which is readily extended to include the Gilbert and Gilbert–Jenkins theories of mass transport of interacting systems.

Fick's Second Law of Diffusion. In all likelihood, the reader has already concluded from the invariance of Fick's second law [Eq. (156)] under the transformation ($\bar{x} = ax$, $\bar{t} = a^2 t$) that $\eta = x/t^{1/2}$ is a similarity variable for this equation. However, the basis for having chosen the particular set of illustrative transformations may not be so apparent. Herein lies the power of the general theory of one-parameter groups of transformations. As per Eq. (158), let the set of transformations be

$$\bar{t} = a^{\alpha_1} t, \qquad \bar{x} = a^{\alpha_2} x, \qquad \bar{c} = a^{\gamma} c \tag{162}$$

in which case

$$dc = \left(\frac{\partial c}{\partial \bar{x}}\right)_{\bar{t}} d\bar{x} + \left(\frac{\partial c}{\partial \bar{t}}\right)_{\bar{x}} d\bar{t} = a^{-\gamma} d\bar{c}$$

$$\left(\frac{\partial c}{\partial t}\right)_x = \left(\frac{\partial c}{\partial \bar{t}}\right)_{\bar{x}} \frac{d\bar{t}}{dt} = \left(\frac{\partial c}{\partial \bar{t}}\right)_{\bar{x}} a^{\alpha_1}$$

$$\left(\frac{\partial c}{\partial x}\right)_t = \left(\frac{\partial c}{\partial \bar{x}}\right)_{\bar{t}} \frac{d\bar{x}}{dx} = \left(\frac{\partial c}{\partial \bar{x}}\right)_{\bar{t}} a^{\alpha_2} \tag{163}$$

$$\left(\frac{\partial^2 c}{\partial x^2}\right)_t = a^{\alpha_2} \left[\frac{\partial}{\partial x}\left(\frac{\partial c}{\partial \bar{x}}\right)_{\bar{t}}\right]_t = \left(\frac{\partial^2 c}{\partial \bar{x}^2}\right)_{\bar{t}} a^{2\alpha_2}$$

Upon substitution into Eq. (156), we obtain

$$a^{\alpha_1 - \gamma} (\partial \bar{c}/\partial \bar{t})_{\bar{x}} = a^{2\alpha_2 - \gamma} D(\partial^2 \bar{c}/\partial \bar{x}^2)_{\bar{t}} \tag{164}$$

Now, in order to satisfy the condition of invariance, the exponential coefficients on either side of the equation must be the same. This requirement leads to the algebraic equation

$$\alpha_1 - \gamma = 2\alpha_2 - \gamma \tag{165}$$

Thus, it is found that $\alpha_1 = 2\alpha_2$ and that γ is arbitrary. Accordingly, γ may be assigned the value of zero.[1] Then, from Eq. (159), the similarity variables are

$$\eta = x/t^{1/2}, \qquad c(\eta) = c(x, t) \tag{166}$$

---

[1] Other values of γ generate similarity solutions which apply to initial conditions that are not physically useful.

The Similarity Transformation

These are introduced into Fick's second law as follows:

$$dc = \frac{dc}{d\eta} d\eta$$

$$\left(\frac{\partial c}{\partial t}\right)_x = \frac{dc}{d\eta}\left(\frac{\partial \eta}{\partial t}\right)_x = -\frac{dc}{d\eta}\frac{\eta}{2t}$$

$$\left(\frac{\partial c}{\partial x}\right)_t = \frac{dc}{d\eta}\left(\frac{\partial \eta}{\partial x}\right)_t = \frac{dc}{d\eta}\frac{1}{t^{1/2}} \qquad (167)$$

$$\left(\frac{\partial^2 c}{\partial x^2}\right)_t = \left[\frac{\partial}{\partial x}\left(\frac{dc}{d\eta}\right)\right]_t \frac{1}{t^{1/2}} = \left[\frac{d^2 c}{d\eta^2}\left(\frac{\partial \eta}{\partial x}\right)_t\right]\frac{1}{t^{1/2}} = \frac{d^2 c}{d\eta^2}\frac{1}{t}$$

Substitution into Eq. (156) yields the second-order ordinary differential equation

$$\tfrac{1}{2}\eta(dc/d\eta) = -D(d^2 c/d\eta^2) \qquad (168)$$

We now inquire as to when this transformation (known as the *Boltzmann transformation*) can be used effectively. Let us consider three cases: diffusion in (1) an infinite domain of x; (2) a semiinfinite domain; and (3) a finite sheet of thickness l. The initial conditions for diffusion in an infinite domain $(-\infty \leq x \leq \infty, 0 \leq t < \infty)$ are

$$\begin{array}{lll} c = 0, & x > 0, & t = 0 \\ c = c^0, & x < 0, & t = 0 \end{array} \qquad (169)$$

and the boundary condition is

$$c = c^0, \qquad x = -\infty, \qquad t > 0 \qquad (170)$$

Under the transformation $\eta = x/t^{1/2}$, the first condition becomes

$$c = 0, \qquad \eta = \infty \qquad (171)$$

and the second and third consolidate into

$$c = c^0, \qquad \eta = -\infty \qquad (172)$$

Since the three original conditions required for unique solution of Eq. (156) consolidate into the two conditions required by Eq. (168) and since the latter are expressible in terms of η alone, the *Boltzmann transformation* can be used to solve this problem.

In the case of diffusion in a semiinfinite domain $(0 \leq x \leq \infty, 0 \leq t < \infty)$,

$$\begin{array}{lll} c = c^0, & x > 0, & t = 0 \\ c = c_2, & x = 0, & t > 0 \\ c = c_1, & x = \infty, & t > 0 \end{array} \qquad (173)$$

which, under the transformation, become

$$c = c^0, \quad \eta = \infty$$
$$c = c_2, \quad \eta = 0 \quad (174)$$
$$c = c_1, \quad \eta = \infty$$

This problem can be solved via the *Boltzmann transformation* only if $c^0 = c_1$, in which case the first and third conditions consolidates to give the required number of boundary conditions.

Finally, this method cannot be used in the case of diffusion in a finite sheet of thickness l with boundary conditions

$$c = c^0, \quad x = 0, \quad x = l \quad (175)$$

since, under the transformation, the second condition becomes

$$c = c^0, \quad \eta = l/t^{1/2} \quad (176)$$

which is not expressible in terms of η alone, but involves t explicitly.

We choose to solve Fick's law of diffusion for the case of an infinite domain because of its direct bearing on the transport problems considered in this book. Equation (168) is integrated by standard procedures to give

$$dc/d\eta = A \exp(-\eta^2/4D) \quad (177)$$

where A is a constant whose value is determined by applying the boundary conditions given in Eqs. (171) and (172):

$$\int_{c^0}^{0} dc = A \int_{-\infty}^{+\infty} \exp(-\eta^2/4D) \, d\eta \quad (178)$$

which yields

$$A = -c^0/2(\pi D)^{1/2} \quad (179)$$

By substituting this expression for A into Eq. (177) and recalling Eqs. (166) and (167), we obtain

$$\left(\frac{\partial c}{\partial x}\right)_t = -\frac{c^0}{2(\pi Dt)^{1/2}} \exp\left[\frac{-x^2}{4Dt}\right] \quad (180)$$

which is the familiar relationship between concentration gradient and position for diffusion from an initially sharp boundary in an infinitely long column. It should be noted that, as a corollary, we have also solved the forced-diffusion equation for ideal electrophoresis of an inherently homogeneous, noninteracting macromolecule [see Eqs. (19)–(21) and accompanying discussion].

The Similarity Transformation

Ideal Transport Assuming Negligible Diffusion. Under the assumption that diffusion is negligible, the conservation equations for ideal electrophoresis of an inherently homogeneous, noninteracting macromolecule [Eq. (19)] and sedimentation, to the rectilinear and constant-field approximations [Eq. (35)], reduce to

$$(\partial c/\partial t)_x = -v(\partial c/\partial x)_t \tag{181}$$

where v is the constant electrophoretic or sedimentation velocity. A specific instance to which this otherwise approximate description might apply with reasonable rigor is the electrophoresis of a highly purified virus of spherical shape. We shall solve this equation by two different methods, to emphasize that many solutions of the problem are unavailable by the method of similarity transformation.

The first method of solution consists of a transformation of position variable to $y = x - vt$. Under this transformation,

$$(\partial c/\partial t)_x = -v(\partial c/\partial y)_t + (\partial c/\partial t)_y$$
$$(\partial c/\partial x)_t = (\partial c/\partial y)_t \tag{182}$$

Substituting into Eq. (181) gives the result

$$(\partial c/\partial t)_y = 0 \tag{183}$$

which tells us that the initial spatial distribution is preserved during transport, i.e., an initial boundary of any shape whatsoever is simply transported along the x axis at velocity v. This conclusion leads to the general solution of Eq. (181),

$$x = S(c) + vt \tag{184}$$

for any initial conditions defined by $x = S(c)$ at $t = 0$. For the particular case of an infinitely sharp initial boundary [$S(c)$ assuming all negative values of x for $c = 0$, all positive values for $c = c^0$, and zero for all c between 0 and c^0; v, positive in the direction solvent $\rightarrow$ solution], the boundary remains infinitely sharp as it is translated along the electrophoresis or sedimentation column at velocity v.

Let us now seek a similarity solution of Eq. (181). Using the general theory of one-parameter groups of transformation, we find the similarity variables to be

$$\eta = x/t, \quad c(\eta) = c(x, t) \tag{185}$$

which transform Eq. (181) into the ordinary differential equation

$$\frac{dc}{d\eta}(\eta - v) = 0 \tag{186}$$

whose solution comes in two parts:

1. If $\eta \neq v$, there is only one way of satisfying the equation—namely, $dc/d\eta = 0$, or c is constant.
2. If $dc/d\eta \neq 0$, there is only one point where $dc/d\eta$ can depart from zero—namely, at $\eta = v$. Moreover, $dc/d\eta$ may be ∞ at this point, since $dc/d\eta$ and $(\eta - v)$ vary independently.

Thus, the similarity solution is an infinitely sharp boundary migrating along the electrophoresis or sedimentation column at velocity v. Now, we must determine the initial conditions for which this solution applies. At $t = 0$, η assumes the following values: for all $x \neq 0$, $\eta = \pm \infty$, and Eq. (186) tells us that dc must be zero; for $x = 0$, $\eta = 0/0$ so that dc may be greater than zero. In other words, the similarity solution applies only to an initial boundary which is infinitely sharp, and solutions for all other initial conditions are unavailable by the given similarity transformation.

The Gilbert Theory

Interest in the physicochemical characterization of macromolecular association–dissociation reactions and elucidation of the forces involved dates from the early ultracentrifugal investigations of Eriksson-Quensel and Svedberg (9) on the hemocyanins and the contemporaneous diffusion measurements of Tiselius and Gross (10) on horse hemoglobin. This interest has been heightened in recent years by the implications of such studies for the quantitative understanding of biological activity and control mechanisms. Moreover, the reversible polymerization of proteins and their dissociation into subunits are among the simplest prototypes of the quaternary architecture and self-assembly characteristics of enzyme complexes, viruses, membranes, and contractile systems. Although velocity sedimentation and molecular-sieve chromatography are peculiarly well adapted to the characterization of such macromolecular interactions, the application of electrophoresis to these problems must not be discounted. In fact, the low-temperature tetramerization of β-lactoglobulin in acidic media was first detected electrophoretically. Comparison of moving-boundary electrophoretic patterns with sedimentation measurements led Ogston and Tilley (11) to conclude that the two peaks shown by the descending pattern at pH 4.65 are due to polymerization of the protein.

Today, β-lactoglobulin is, perhaps, one of the best understood of associating–dissociating systems due, in large part, to the elegant physicochemical investigations of Timasheff and his co-workers (12–18) on the three genetic variants: β-lactoglobulins A, B, and C (abbreviated as β-A, β-B, and β-C). In fact, this work is a model for studies on interacting systems, being noteworthy for its logical development: (1) the detection of interaction by electrophoretic and ultracentrifugal analyses, followed sequentially by (2) determination of experimental conditions for maximal interaction, (3) investigation of the kinetics of the reactions, (4) application of Gilbert's theory, and finally, (5) extension and confirmation of the latter result by light scattering. Since we will have occasion to refer to the β-lactoglobulin system from time to time, a brief summary of its pertinent features follows: The three genetic variants of β-lactoglobulin (kinetic molecular weight of 36,000 at isoelectric pH) differ only slightly in their primary structure. Thus, two aspartic acid residues in the amino acid sequence of β-A are replaced by two glycines in β-B, and one glutamic acid (or glutamine) in β-B is replaced by histidine in β-C. Both β-A and β-B dissociate reversibly into two subunits of equal molecular weight when exposed to pH 1.6. In contrast, at pH 4.65 and sufficiently low temperature, both variants associate reversibly to form a tetramer, which, in the case of β-A, at least, is of closed structure with 422 symmetry. The association of β-B is much weaker than for β-A, and the two variants form mixed molecular aggregates of intermediate strength. The difference in the free energies of the A–A, B–B, and A–B bonds is due to a difference in entropy, which, in turn, may reflect the small difference in primary structure of β-A and β-B.

Moving-boundary transport experiments on reversibly associating–dissociating macromolecules present no interpretative difficulties if the rates of reaction are sufficiently slow. In that event, the time required for separation of the various species is short compared to the time for significant reequilibration to occur. Accordingly, the transport patterns will resolve into two or more boundaries corresponding to reactants and products; conventional analysis of the patterns yields their mobilities and concentrations in the initial equilibrium mixture. An example is provided by sedimentation (19) and electrophoretic (20) experiments on the reversible dissociation of the globulin arachin. The situation is considerably more complex, however, when rapid chemical reequilibration occurs between species during their differential transport. In this case, peaks in the transport patterns cannot be placed into correspondence with individual reactants and products, and an entirely different method of analysis of the patterns is required. Examples include the above-mentioned tetramerization of β-lactoglobulin A and the polymerization of α-chymotrypsin (21) at appropriate pH and ionic strength. Concern about the interpretation of the sedimentation behavior of the latter protein prompted Gilbert (1, 2) to formulate his theory of sedimentation and electrophoresis of reversibly polymerizing systems described below.

The simplest model of reversible polymerization is one in which monomer coexists with a single polymer,

$$n\mathrm{M} \rightleftarrows \mathrm{P} \tag{187}$$

where M and P represent monomer and polymer, and n is the degree of polymerization. The corresponding set of transport-interconversion equations under the assumption of negligible diffusion is

$$(\partial c_M/\partial t)_x = -v_M(\partial c_M/\partial x)_t + R_M \qquad (188)$$

$$(\partial c_P/\partial t)_x = -v_P(\partial c_P/\partial x)_t - R_M \qquad (189)$$

in which, for sake of simplicity, the concentrations are expressed as grams per liter. Addition of this pair of equations yields

$$(\partial c_M/\partial t)_x + (\partial c_P/\partial t)_x = -v_M(\partial c_M/\partial x)_t - v_P(\partial c_P/\partial x)_t \qquad (190)$$

which is a statement of conservation of total protein during differential transport of monomer and polymer molecules. A second equation must express the effect of the association–dissociation reaction on the concentrations. Gilbert's theory is for systems characterized by reaction rates sufficiently large compared to the rate of separation of the two species in the external field that local equilibrium obtains at every instant. Accordingly,

$$K' = c_P/c_M^n \qquad (191)$$

where K' is the equilibrium constant for reaction (187). Elimination of c_P between Eqs. (190) and (191) yields the nonlinear partial differential equation

$$(\partial c_M/\partial t)_x[1 + nK'c_n^{n-1}] = -[v_M + v_P nK'c_M^{n-1}](\partial c_M/\partial x)_t \qquad (192)$$

for which we seek a similarity solution. As per Eq. (158), consider the set of transformations

$$\bar{t} = a^{\alpha_1}t, \qquad \bar{x} = a^{\alpha_2}x, \qquad \bar{c}_M = a^\gamma c_M \qquad (193)$$

which transform Eq. (192) into

$$\left(\frac{\partial \bar{c}_M}{\partial \bar{t}}\right)_{\bar{x}} a^{\alpha_1 - \gamma}[1 + nK'\bar{c}_M^{n-1}a^{-\gamma(n-1)}]$$
$$= -[v_M + v_P nK'\bar{c}_M^{n-1}a^{-\gamma(n-1)}]\left(\frac{\partial \bar{c}_M}{\partial \bar{x}}\right)_{\bar{t}} a^{\alpha_2 - \gamma} \qquad (194)$$

The condition of invariance requires that

$$\alpha_1 - \gamma = \alpha_1 - n\gamma = \alpha_2 - \gamma = \alpha_2 - n\gamma \qquad (195)$$

which is satisfied by $\alpha_1 = \alpha_2$, $\gamma = 0$. The similarity variables defined by Eq. (159) are

$$\eta = x/t, \qquad c_M(\eta) = c_M(x, t) \qquad (196)$$

which transform Eq. (192) into the nonlinear ordinary differential equation

$$\frac{dc_M}{d\eta}\eta = \frac{v_M + v_P nK'c_M^{n-1}}{1 + nK'c_M^{n-1}}\frac{dc_M}{d\eta} \qquad (197)$$

The Gilbert Theory

There are two solutions to this equation, the discontinuous solution of Wilson (22) and the continuous one of Gilbert (1); however, no ambiguity arises, since physical considerations indicate which solution is the stable one for a given set of conditions.[2] We shall proceed by first obtaining each of the solutions and then applying them to specific situations.

Wilson's Discontinuous Solution. We note that Eq. (197) can be recast in the form

$$\frac{dc_M}{d\eta}\eta = \left[\bar{v} + \frac{c_M + K'c_M^n}{1 + nK'c_M^{n-1}}\frac{d\bar{v}}{dc_M}\right]\frac{dc_M}{d\eta} \quad (198)$$

where $\bar{v}$ is the local weight-average velocity defined by

$$\bar{v} = (v_M + v_P K'c_M^{n-1})/(1 + K'c_M^{n-1}) \quad (199)$$

If $\bar{v}$ is the same at every position in the transport column where there is macromolecule, $d\bar{v}/dc_M = 0$, and Eq. (198) reduces to

$$(dc_M/d\eta)\eta = \bar{v}^0(dc_M/d\eta) \quad (200)$$

in which $\bar{v}^0$ is the constituent velocity of the original equilibrium mixture in the plateau region ahead of a sedimenting or descending electrophoretic boundary or behind an ascending electrophoretic boundary. This equation is of the same form as Eq. (186), and admits only the discontinuous solution

$$\eta \neq \bar{v}^0, \quad \frac{dc_M}{d\eta} = 0; \quad \frac{dc_M}{d\eta} \neq 0, \quad \eta = \bar{v}^0 \quad (201)$$

which applies only to an infinitely sharp initial boundary.

Gilbert's Continuous Solution. If the value of $\bar{v}$ is dependent upon position, Eq. (197) admits a continuous solution which comes in two parts:

$$dc_M/d\eta = 0, \quad \text{or} \quad c_M \text{ is constant} \quad (202)$$

$$dc_M/d\eta \neq 0, \quad \eta = (v_M + v_P K'c_M^{n-1})/(1 + nK'c_M^{n-1}) \quad (203)$$

---

[2] The differential equation of chromatography is of the same form as Eq. (192). In one of the earliest treatments of the theory of chromatography, Wilson (22) found that the differential equation admits a discontinuous solution for initial and boundary conditions corresponding to an initial solute zone having infinitely sharp leading and trailing edges and, effectively, to the condition that, during elution, the weight average rate of migration of solute is everywhere the same. When applied to both edges, this solution describes a developing zone which maintains constant width and sharply defined edges as it migrates through the chromatographic column. This is, of course, an inaccurate description (23–25) of the actual situation, because the discontinuous solution was applied without regard for its instability under the conditions which obtain at one or the other edge of the zone. Nevertheless, the fact remains that the differential equation does admit a discontinuous solution.

The first part, Eq. (202), applies to the plateau and the region of the cell devoid of macromolecule, and the second part, Eq. (203), applies to the boundary separating the two. The two parts must be joined at two values of η—that value for which c_M is equal to the concentration in the plateau, and the value for which $c_M = 0$—to give a solution which applies everywhere in the transport column.

To answer questions as to initial conditions and convergence, we rearrange Eq. (197) to read

$$\frac{dc_M}{d\eta}[\eta - \lambda(c_M)] = 0 \tag{204}$$

where $\lambda(c_M)$ is the variable coefficient of the derivative on the right-hand side of Eq. (197), and note that differentiation of Eq. (203) yields

$$\left[\frac{dc_M}{d\eta}\right]^{-1} = \frac{(\eta - v_P)^2}{(v_P - v_M)}(n-1)[nK']^{[(n-3)/(n-1)]}c_M^{n-2}$$
$$= \frac{(v_M - \eta)^2}{(v_P - v_M)}(n-1)[nK']^{-[(n+1)/(n-1)]}c_M^{-n} \tag{205}$$

Now, at $t = 0$, $\eta = \infty$ for all $x \neq 0$, and, from Eq. (204), $dc_M/d\eta$ must be zero; for $x = 0$, $\eta = 0/0$, and $dc_M/d\eta$ may be greater than zero. In other words, the similarity solution applies only to an infinitely sharp initial boundary.

As for convergence of the differential equation when $dc_M/d\eta \to \infty$, we inquire whether

$$\lim_{dc_M/d\eta \to \infty} \frac{[\eta - \lambda(c_M)]}{[dc_M/d\eta]^{-1}} = 0 \tag{206}$$

Equations (203) and (205) tell us that $dc_M/d\eta$ can be infinite at two positions:

1. For $n \geq 3$, as $\eta \to v_M$, $c_M \to 0$ and $dc_M/d\eta \to \infty$. Since $[\eta - \lambda(c_M)] \to 0$ as c_M^{n-1} and $[dc_M/d\eta]^{-1} \to 0$ as c_M^{n-2}, the numerator of Eq. (206) approaches zero more rapidly than does the denominator. Accordingly, $\lim_{\eta \to v_M} = 0$.

2. For all n, as $\eta \to v_P$, $c_M \to \infty$ and $dc_M/d\eta \to \infty$. Since, in this case, $[\eta - \lambda(c_M)] \to 0$ as $1/c_M^{n-1}$, and $[dc_M/d\eta]^{-1} \to 0$ as $1/c_M^n$, $\lim_{\eta \to v_P} = \infty$. However, this is of no consequence, since we do not take the solution all the way to this point, i.e., Eqs. (202) and (203) join at $\eta < v_P$.

Sedimentation. For sedimentation of a reversibly polymerizing system, Wilson's discontinuous solution is unstable, since polymeric molecules in the front tend to dissociate into monomeric molecules, which lag behind. Thus, an infinitely sharp front cannot be maintained, and Gilbert's solution develops naturally. Now, from Eqs. (203) and (191), we find that

$$c_M = \left(\frac{1}{nK'}\right)^{1/(n-1)}\left(\frac{v_M - \eta}{\eta - v_P}\right)^{1/(n-1)} \tag{207}$$

and

$$c_P = K'c_M^n = K'\left[\frac{1}{nK'}\left(\frac{v_M - \eta}{\eta - v_P}\right)\right]^{n/(n-1)} \quad (208)$$

so that the total concentration of protein as a function of the similarity variable η is

$$c_M + c_P = \left(\frac{1}{nK'}\right)^{1/(n-1)}\left[\frac{v_M - \eta}{\eta - v_P}\right]^{1/(n-1)}\left[1 + \frac{1}{n}\left(\frac{v_M - \eta}{\eta - v_P}\right)\right] \quad (209)$$

This relationship illustrates the physical meaning of the term "similarity" quite nicely. If we were to compute the integral sedimentation pattern, $(c_M + c_P)$ vs. x, for any time of sedimentation t, the pattern at any other time t_1 could be obtained therefrom simply by multiplying the abscissa by the scaling factor t_1/t. On the other hand, all stages of the transport process can be visualized in a single display of $(c_M + c_P)$ vs. η. It is convenient, however, to make the coordinate transformation

$$\delta = (\eta - v_M)/(v_P - v_M) \quad (210)$$

In so doing, we adopt a description in which motion of the polymer in the boundary is with respect to stationary monomer. Under this transformation, Eq. (209) becomes

$$c_M + c_P = \left(\frac{1}{nK'}\frac{\delta}{1-\delta}\right)^{1/(n-1)}\left[1 + \frac{1}{n}\frac{\delta}{1-\delta}\right] \quad (211)$$

Since the sedimentation pattern is usually a photographic record of concentration gradient versus position, we differentiate Eq. (211) with respect to δ:

$$\frac{d(c_M + c_P)}{d\delta} = \frac{1}{(n-1)}\left(\frac{1}{nK'}\right)^{1/(n-1)}\delta^{(2-n)/(n-1)}\left(\frac{1}{1-\delta}\right)^{(2n-1)/(n-1)} \quad (212)$$

or, if one so chooses,

$$\left[\frac{\partial(c_M + c_P)}{\partial x}\right]_t = \frac{d(c_M + c_P)}{d\delta}\left(\frac{\partial\delta}{\partial x}\right)_t = \frac{d(c_M + c_P)}{d\delta}\frac{1}{(v_P - v_M)t} \quad (213)$$

Now, for $\eta = v_M$, $\delta = 0$ and $c_M + c_P = 0$, i.e., the trailing edge of the boundary moves at the velocity of the monomer. For $\eta = v_P$, $\delta = 1$ and $c_M + c_P = \infty$, which means that the leading edge of the boundary must move less rapidly than the polymer and that $0 \leq \delta < 1$. The upper value of δ is that value for which $c_M + c_P$ equals the total protein concentration in the plateau region of the sedimentation column. The two parts [Eqs. (202) and (203)] of Gilbert's solution join here and also at $\delta = 0$, $c_M + c_P = 0$, to give a solution everywhere in the centrifuge cell. In practice, both $c_M + c_P$ and $d(c_M + c_P)/d\delta$ are computed for increasing values of δ. At that value for which the computed concentration becomes equal to the plateau concentration, computation is halted and the concentration gradient returned to zero.

Underlying the foregoing discussion is the assumption of instantaneous reestablishment of chemical equilibrium during differential transport of monomer and polymer. In that event, monomer and polymer must always coexist at all positions in the centrifuge cell where the concentration of macromolecule is finite. It does not necessarily follow, however, that the sedimenting boundary will be unimodal. As Gilbert (2) has pointed out, partial resolution of a reaction boundary into peaks "merely requires the appropriate concentration gradients to be present in a single continuous

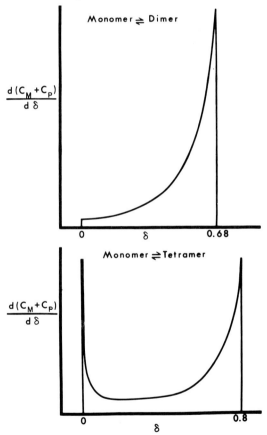

FIG. 30. Comparison of theoretical sedimentation-velocity patterns for dimerizing and tetramerizing systems, $nM \rightleftharpoons P$. Total macromolecular concentration in the plateau region, 27.8 g liter^{-1}; 50% by weight of the single polymer. Sedimentation is from left to right.

boundary and there is no physical reason why this should not be so." It is logical, therefore, to examine the theoretical gradient curve for minima by differentiating Eq. (212) with respect to δ and equating to zero. When this is done, a single minimum is found at

$$\delta_{\min} = (n-2)/3(n-1) \tag{214}$$

This result expresses the most dramatic prediction of the Gilbert theory. Whereas only a single sedimenting peak will be observed for macromolecular dimerization ($n = 2$), for higher-order polymerization reactions ($n \geq 3$) the reaction boundary may resolve into two peaks even though reequilibration is instantaneous. This important prediction is exhibited diagrammatically in Fig. 30, which contrasts theoretical sedimentation patterns for dimerizing and tetramerizing systems. It must be emphasized that neither of the peaks in the latter pattern can be identified with a distinct macromolecular entity, since both monomer and polymer coexist in equilibrium at every position where the concentration is finite.

As illustrated in Fig. 31, resolution of the reaction boundary into two peaks can occur for $n \geq 3$ only when the total macromolecular concentration in the plateau region exceeds the concentration Δ_s, corresponding to $\delta_{\min}$. At lower concentrations, the single peak grows in area as the concentration is increased to the value Δ_s. Upon increasing the concentration still further, a second, more rapidly sedimenting peak appears and grows in area, while that of the slower peak now remains unchanged. This constancy of the area of the slower

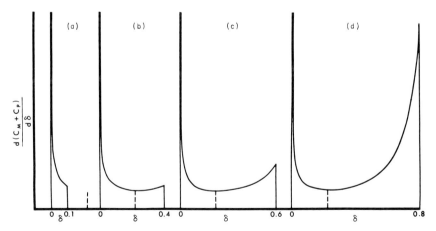

FIG. 31. Effect of total macromolecular concentration on the theoretical sedimentation-velocity pattern of a polymerizing system in which monomer coexists in equilibrium with tetramer alone: (a) 4.51 g liter^{-1}; (b) 9.17 g liter^{-1}; (c) 14.2 g liter^{-1}; (d) 28.6 g liter^{-1}. Vertical line locates $\delta_{\min}$; $\Delta_s = 6.35$ g liter^{-1}; $K' = 3.43 \times 10^{-4}$ liter^3 g^{-3}. Sedimentation is from left to right.

peak with increasing total macromolecule concentration is diagnostic for higher-order polymerization reactions.

It is also apparent from Fig. 31 that the sedimentation velocity of the single peak observed at concentrations less than Δ_s increases with increasing concentration from v_M at infinite dilution to a slightly larger value at Δ_s due to the simultaneous presence of some polymer. The velocity of the slower peak in the emergent bimodal boundary has the latter value independent of any further increase in concentration. In contrast, the velocity of the faster peak increases with increasing concentration, but is always less than for pure polymer. In fact, there is no direct way of determining the velocity of the polymer from the sedimentation patterns. This is so because there is no position in the boundary corresponding to polymer, i.e., $0 \leq \delta < 1$. Thus, the first moment of the boundary (square root of the second moment for a sector-shaped cell) gives the weight-average velocity in the plateau region.[3] The weight-average value decreases with decreasing concentration and extrapolates at infinite dilution to the velocity of the monomer. The sedimentation velocity of the polymer can be estimated, however, using an indirect method suggested by Nichol and Bethune (26), who noted that Eq. (214) may be written as

$$\frac{x_1 - x_2}{x_3 - x_2} = \frac{n-2}{3(n-1)} \tag{215}$$

where x_1, $x_2 = v_M t$, and $x_3 = v_P t$ are the distances from the meniscus to the positions of the minimum in the reaction boundary, the position of the

---

[3] That the first moment of the boundary $\bar{\delta}$ gives the weight average sedimentation velocity $\bar{v}^0$ in the plateau region of a rectilinear sedimentation cell can be seen as follows:

$$\bar{\delta} = \frac{\int_0^{c_M{}^0 + c_P{}^0} \delta \, d(c_M + c_P)}{\int_0^{c_M{}^0 + c_P{}^0} d(c_M + c_P)} = \frac{1}{c_M{}^0 + c_P{}^0} \int_0^{c_M{}^0 + c_P{}^0} \left(\frac{\eta - v_M}{v_P - v_M}\right) d(c_M + c_P)$$

where $c_M{}^0$ and $c_P{}^0$ are the concentrations of monomer and polymer in the plateau. By substituting Eq. (203) for η and recalling that $d(c_M + c_P) = (1 + nK'c_M{}^{n-1}) \, dc_M$, one obtains

$$\bar{\delta} = [1/(c_M{}^0 + c_P{}^0)(v_P - v_M)] \int_0^{c_M{}^0} (v_M + v_P n K' c_M{}^{n-1}) \, dc_M - [v_M/(v_P - v_M)]$$
$$= (\bar{v}^0 - v_M)/(v_P - v_M)$$

By using the exact conservation equation for the sedimentation of a polymerizing system [Eq. (38) in Chapter I with $\bar{c}$, $\tilde{D}$, and $\bar{s}$ replaced by the constituent quantities $c_M + c_P$, $\overline{D}$ and $\bar{s}$] and proceeding as in the derivation of Eq. (42) in Chapter 1, it is found that, for a sector-shaped cell, the constituent (weight-average) sedimentation coefficient in the plateau region is given by the logarithmic rate of movement of the square root of the second moment of the reaction boundary rather than the first moment.

monomer, and the hypothetical position of the polymer if it were to exist alone. The position of the monomer is rigorously the very trailing edge of the boundary ($\delta = 0$), but in practice is taken as the position of the slower peak. For a given value of n and measured values of x_1 and x_2, values of x_3 may be found as a function of time. Calculation of the sedimentation constant of the polymer follows directly.

Implicit in the above considerations is the fact that Δ_s is a characteristic parameter of polymerizing systems, being a function only of n and K'. By substituting Δ_s for $c_M + c_P$ and the expression for δ_{min} [Eq. (214)] for δ in Eq. (211), one obtains, upon rearrangement, the relationship

$$K' = \Delta_s^{1-n} [2(n^2-1)]^{n-1} (n-2)/[n(2n-1)]^n \qquad (216)$$

Thus, to determine K', it is only necessary to determine Δ_s from the area of the slower peak in the bimodal reaction boundary obtained at any concentration for which the upper limit of δ exceeds δ_{min}. This presupposes, of course, that the value of n has been determined from some independent physicochemical measurement on the system. It is interesting to note that Fujita (27) has developed the Gilbert theory for sedimentation in a sector-shaped cell. Although deviations of his theoretical patterns from these described above for a rectilinear cell are minor in view of the neglect of diffusion in both cases, one consequence (28) of his computations is that Eq. (216) represents a first approximation of the case where radial dilution is taken into consideration. In practice, values of K' computed from Eq. (216) will drift with concentration and must be extrapolated to infinite dilution.

As we have seen above, polymerizing systems of the type under consideration fall into two classes, depending upon whether $n = 2$ or $n \geq 3$. Examples of the first class include the reversible dissociation into half-molecules of hemoglobin (29) at high salt concentrations and β-lactoglobulin A and B (15) at very low pH (NaCl–HCl, ionic strength 0.3, pH 1.6, 2–25°C). The sedimentation patterns of such systems characteristically show a single peak whose sedimentation coefficient exhibits a distinctive dependence upon concentration (Fig. 32a). The sedimentation coefficient may at first increase with decreasing concentration as a consequence of hydrodynamic effects, then pass through a maximum due to progressive molecular dissociation under the influence of mass action, and then extrapolate to the sedimentation coefficient of the subunit or monomeric molecule. The distinctive shape of the plot of sedimentation coefficient versus concentration provides a means for detecting interactions[4] which is particularly valuable when bimodality is not observed.

---

[4] A decrease in sedimentation coefficient with decreasing concentration is not unique for associating–dissociating systems, however, and could conceivably arise from suitable variation in frictional coefficient or partial specific volume with concentration, as, for example, with highly charged flexible molecules (30).

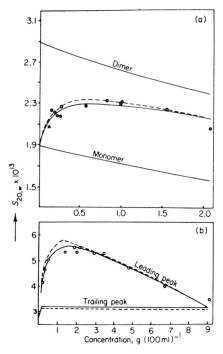

FIG. 32. Concentration dependence of the sedimentation coefficient of β-lactoglobulin: (a) molecular dissociation of β-lactoglobulin A and B at pH 1.6 (15); (b) low-temperature tetramerization of β-lactoglobulin A at pH 4.65 (13). Theoretical curves: (———) weight-average sedimentation cofficient; (- - -) "median" sedimentation coefficient. Values of the parameters used in the calculation of theoretical curves are as follows: (a) $K' = 0.4 \times 10^4$ liters mole^{-1}, $n = 2$, $v_M^\infty = 1.89$ S, $v_P^\infty = 2.87$ S; (b) $K' = 5 \times 10^{11}$ liter^3 mole^{-3}, $n = 4$, $v_M\infty = 2.87$ S, $v_P^\infty = 7.23$ S. [Taken from Nichol and co-workers (28), who, in turn, adapted figures of Gilbert and Gilbert (32, 33).]

However, it must be borne in mind that extrapolation to infinite dilution of the linear portion of the curve sometimes observed on the high-concentration side of the maximum does not give the sedimentation coefficient of the dimer (31, 32). This is manifest in the case of β-lactoglobulin by the data presented in Fig. 32a. Interpretation of these data in terms of a simple dissociation of the protein into two subunits of equal molecular weight depends, of course, upon the correlation of velocity sedimentation experiments with independent physicochemical measurements, in this case, light-scattering measurements and molecular-weight determinations using the Archibald method. The two theoretical curves passing through the data points were obtained (32) by computing either weight-average sedimentation coefficients or "median" sedimentation coefficients both as functions of concentration. The "median"

The Gilbert Theory

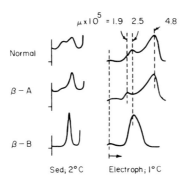

FIG. 33. Sedimentation and descending moving-boundary electrophoretic patterns of β-lactoglobulin under optimal conditions for tetramerization, 0.1 ionic strength acetate buffer, pH 4.65, 1–2°C. "Normal" β-lactoglobulin is a mixture of the two genetic variants β-A and β-B. Sedimentation: 59,780 rpm; "normal" and β-A, 14 g liter^{-1}, 160 min; β-B, 70 g liter^{-1}, 352 min. Electrophoresis: 16 g liter^{-1}, 8000 sec at 9.7 V cm^{-1}. [From Brown and Timasheff (38).]

sedimentation coefficient is given by the value of δ in Eq. (211) for which $c_M + c_P$ is one-half the plateau concentration. The following parametric assignments were made for both sets of calculations: (1) values of n and K' from light-scattering and Archibald measurements; (2) s_M^∞ by extrapolation of observed sedimentation coefficients to infinite dilution; (3) $s_P^\infty = s_M^\infty \times (n^{2/3}/1.044)$, in which $n^{2/3}$ and $1/1.044$ account for the differences in weight and shape of monomer and polymer; and (4) g in the usual hydrodynamic relationship $s = s^\infty (1 - gc)$ for each species considered alone, from the slope of the approximately linear portion of the observed sedimentation coefficient–concentration curve at high concentration. The strikingly good agreement between the theoretical curves and the data points inspires confidence in both the experimental approach and the Gilbert theory.

The most thoroughly characterized example of an associating system for which $n \geq 3$ is the low-temperature tetramerization (13, 14, 17) of β-lactoglobulin A (acetate buffer, ionic strength 0.1, optimal pH 4.40–4.65, 1–4.5°C). The sedimentation pattern is bimodal (Fig. 33); the area of the slower peak remains constant over a threefold range of concentration[5] and the sedimentation coefficient of the leading peak exhibits the concentration dependence (Fig. 32b) described above as characteristic of rapidly reequilibrating association–dissociation reactions. A companion light-scattering investigation

---

[5] Initial observations on the tetramerization of β-lactoglobulin A suggested that Δ_s might increase with increasing total concentration. Subsequently, it was shown, however, that the increasing area of the slower peak was due to the presence of a small amount of inert material, and that Δ_s is, in fact, constant for very highly purified β-lactoglobulin A (17).

showed that intermediate-size polymers are present in negligible amounts, so that monomer may be regarded as coexisting in equilibrium with tetramer alone. These measurements also showed that reequilibration is rapid. Moreover, the value of K' computed from the sedimentation patterns using Eq. (216) is in reasonably good agreement with the value from light-scattering measurements.[6] The theoretical curves passing through the data points in the sedimentation coefficient–concentration plot presented in Fig. 32b were computed (33) essentially as described above for the dissociation of β-lactoglobulin into subunits. Once again, the agreement between theory and experiment is impressive and demonstrates the applicability of the Gilbert theory to this system.

Before turning to other matters, a few words concerning the sedimentation behavior of mercuripapain and β-lactoglobulin B are in order as the results of these experiments bring into focus some of the precautions which must be exercised when interpreting sedimentation patterns of associating systems. Papain forms a crystalline mercury derivative which contains 1 mole of mercury per 2 moles of protein and whose sedimentation behavior depends upon pH (34). While the sedimentation patterns show a single symmetrical peak with the same sedimentation coefficient as papain at pH 4,[7] two peaks may be obtained at pH 8, depending upon the concentration of mercuripapain. A set of sedimentation patterns obtained at progressively increasing concentration is displayed in Fig. 34. It is immediately apparent that the concentration dependence of the patterns is virtually the same as that predicted by the Gilbert theory for higher-order polymerization (Fig. 31). Thus, essentially a single peak with about the same sedimentation coefficient as papain is observed at the lowest concentration, but, as the concentration is increased, a considerably faster peak appears and grows in area, while that of the slower peak remains reasonably constant. Clearly at pH 8, mercuripapain exists in solution as an equilibrium mixture of monomer with *at least* one kind of polymer for which $n > 2$. But the point which we wish to emphasize here is that, if the apparent single peak at the lowest concentration were an isolated observation, one might erroneously conclude that a single species

---

[6] Values of K' estimated from sedimentation (13) and electrophoresis (40) at 1–2°C are 2.7×10^{-2} and 1.5×10^{-2} liter^3 g^{-3}, respectively. The value derived from light-scattering data depends upon the description adopted: for monomer $\rightleftharpoons$ tetramer (14), 1.7×10^{-2} at 4.5°C; for monomer $\rightleftharpoons$ dimer $\rightleftharpoons$ trimer $\rightleftharpoons$ tetramer (17), 3.2×10^{-2} at 4.5°C and 11×10^{-2} at 2°C. While the latter is the more exact description, the simple monomer–tetramer model is a good approximation at temperatures close to 0°C and also at 4.5°C for concentration greater than 15 g liter^{-1}.

[7] This is a somewhat unusual case of protein–mercury ion interaction, in that one might anticipate a dimer complex as in the case of mercaptoalbumin (35). Although ultracentrifugal analysis at pH 4 does not indicate any change in size, papain does remain combined with mercury at this pH as revealed by electrophoretic measurements.

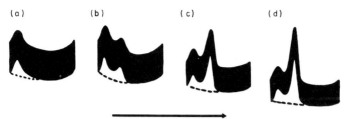

FIG. 34. Sedimentation patterns of crystalline mercuripapain in ionic strength 0.1 tris (hydroxymethyl) aminomethane buffer, pH 8. Each pattern was obtained in a separate experiment performed with a different total protein concentration: (a) 3.3 g liter^{-1}; (b) 6.5 g liter^{-1}; (c) 13 g liter^{-1}; (d) 22.8 g liter^{-1}. Sedimentation is from left to right. [Adapted from Smith and his co-workers (34).] Values of Δ_s (28): (a) 3.2 g liter^{-1}; (b) 3.7 g liter^{-1}; (c) 3.5 g liter^{-1}; (d) 4.8 g liter^{-1}.

exists in solution or that mercuripapain dimerizes. Obviously, sedimentation experiments at different concentrations are essential for the interpretation of associating systems.

Even then, one may be led astray. Consider, for example, the low-temperature association of β-lactoglobulin B (17) under conditions identical to those which are optimal for the tetramerization of β-lactoglobulin A. The fact that the sedimentation patterns of the former show a single peak over a wide range of concentration (Fig. 33) understandably gave rise to the notion that this genetic variant cannot form aggregates larger than a dimer. Subsequently, however, highly precise light-scattering and sedimentation measurements revealed that the protein actually tetramerizes weakly with a value of K' an order of magnitude lower than for β-lactoglobulin A. Thus, the difference between the two genetic variants is a quantitative one and not one of kind as far as their ability to associate is concerned. Now, failure of β-lactoglobulin B to develop bimodal reaction boundaries evidently resides in the interplay between the strength of association and the hydrodynamic concentration dependence of the sedimentation coefficients. As we have seen, formulation of the Gilbert theory makes the simplifying assumption of constant sedimentation velocities. Gilbert (36) subsequently showed that, while this assumption is of little consequence for strong associations, it cannot be made for weak ones. In fact, for sufficiently weak association, the interplay between the strength of association and hydrodynamic retardation of sedimentation of the species in equilibrium assumes such proportion as to overwhelm the tendency for bimodality to develop. One way of viewing the situation is that Δ_s is so extremely large that the expected bimodality never develops as the two peaks are hydrodynamically merged into one. Finally, these results underscore the fact that unambiguous interpretation of sedimentation patterns of reversibly polymerizing systems relies heavily on independent physicochemical measurements and that mere curve fitting of sedimentation-coefficient–

concentration data to obtain unique values of the several parameters is not justified.

Above, we have been concerned with systems in which monomer coexists in equilibrium with a single polymer. For higher-order polymerizations, however, one might expect intermediate-size polymers to exist in significant concentrations, depending upon the value of the association constants. Thus, for example, the ionic-strength-dependent association of α-chymotrypsin at pH 7.9 apparently exhibits a transitional behavior on going from ionic strength 0.03 to 0.3. While the sedimentation pattern is bimodal at ionic strength 0.03, a single peak is observed at the higher ionic strength even though association still occurs. Various lines of evidence indicate that a single higher-order polymer predominates at the lower ionic strength while intermediates are present in comparable concentrations at the higher ionic strength. The reader is referred to Nichol and his co-workers (28) and to Gilbert (1, 2) for discussions of this interesting problem and pertinent literature citations. A similar situation obtains in the temperature-dependent tetramerization of β-lactoglobulin A at pH 4.65. As we have seen, this system may be regarded as one in which, at 1–4.5°C, monomer coexists in equilibrium with tetramer alone. In contrast (17), at 8°C, the protein undergoes a weaker, progressive tetramerization of the type monomer ⇌ dimer ⇌ trimer ⇌ tetramer. The concentration distribution of material among the four species is presented in Fig. 35 as a function of total concentration. Clearly, extension of the Gilbert theory to include situations in which intermediate polymers exist in significant concentrations is of considerable importance.

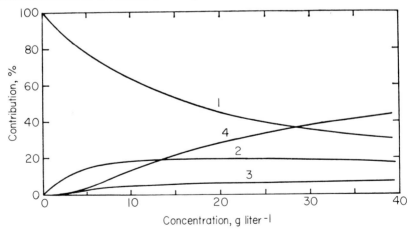

FIG. 35. Progressive tetramerization (monomer ⇌ dimer ⇌ trimer ⇌ tetramer) of β-lactoglobulin A at pH 4.65 and 8°C. Percent contribution of each species to the total protein concentration as a function of total concentration: (1) monomer; (2) dimer; (3) trimer; (4) tetramer. [From Kumosinski and Timasheff (17).]

Gilbert (2, 36) has extended his two-species theory to include progressive polymerization. The formulation includes not only the existence of intermediate polymers, but also allows for the possibility of more than one form of the n-mers, e.g., linear and cyclic structures. Although the general solution is not in closed form, Rao and Kegeles (37) have obtained a closed-form solution for the coexistence of monomer, dimer, and trimer. The predictions of the general theory can be summarized as follows: The sedimentation pattern will show a single asymmetrical peak when the concentration of the highest polymer is proportionally low. As the concentration of the highest polymer increases relative to the concentrations of the intermediates, a shoulder develops on the trailing edge of the peak. Upon further increase in the relative concentration of the highest polymer, bimodality develops; finally, in the limit where the concentrations of intermediates become vanishingly small, the reaction boundary assumes the bimodal form predicted by the simple two-species theory. Finally, Bethune and Grillo (37a) have shown that the system, monomer $\rightleftharpoons$ trimer $\rightleftharpoons$ nonamer, can give reaction boundaries showing one, two, or three peaks, depending upon the value of the equilibrium constants and the total macromolecular concentration. Experimental criteria are given for defining a system as one in which these reactions occur.

Electrophoresis. One's first thought concerning the relationship of electrophoretic mobility to the state of molecular association of a protein is likely to be that the greater the extent of association, the lower is the mobility. Upon reflection, however, it becomes apparent that the situation is not nearly so simple. Thus, for example, it is possible that the increase in size and, perhaps, asymmetry may be offset by an increase in zeta potential for one reason or another. Although it is not generally possible to predict *a priori* the relationship of the electrophoretic mobility of an aggregated protein molecule to that of the monomer, it is clear that changes in the state of aggregation sometimes do not influence the mobility significantly—for example, acid-modified conalbumin (39). In other cases, such as the low-temperature tetramerization of β-lactoglobulin A, aggregation causes the mobility to increase (13).

When the monomer is in rapid equilibrium with a single polymer of greater mobility, the Gilbert theory of sedimentation is equally applicable to descending moving-boundary electrophoresis. This is so since his mathematical equations make the rectilinear approximation and can be applied directly to electrophoresis simply by substitution of electrophoretic velocities for sedimentation velocities. Accordingly, it is anticipated that the descending electrophoretic pattern will show a single migrating peak for $n = 2$ and a bimodal reaction boundary for $n \geq 3$, provided, of course, that the total macromolecular concentration exceeds the characteristic value Δ_s. Once again, the tetramerization of β-lactoglobulin A provides an excellent experimental test

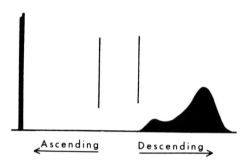

Fig. 36. Moving-boundary electrophoretic patterns of very highly purified β-lactoglobulin A under optimal conditions for tetramerization, 0.1 ionic strength acetate buffer, pH 4.66, at 1°C. Protein concentration, 15.8 g liter^{-1}; position of the initial boundary indicated by vertical lines. [Adapted from Tombs (40).]

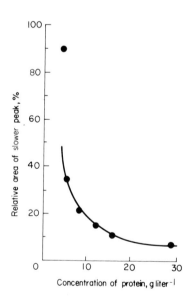

Fig. 37. Composition of the descending electrophoretic boundary of tetramerizing β-lactoglobulin A: variation of the *relative* area of the slower peak of the bimodal reaction boundary with total protein concentration. Theoretical curve computed from the general equation (independent of n and K'). Percent slow peak = $\frac{1}{2}${(total concentration at which area of slow peak is 50% of total area)/(total concentration)} × 100 = [Δ_s/(total concentration)] × 100. [Adapted from Tombs (40).]

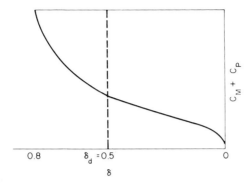

FIG. 38. Theoretical ascending moving boundary for a polymerizing system in which monomer coexists in equilibrium with tetramer alone. Integral plot of concentration $c_M + c_P$, against position variable δ: (——) Gilbert's continuous solution; (- - -) Wilson's discontinuous solution. Total protein concentration, 28.6 g liter^{-1}; $K' = 3.43 \times 10^{-4}$ liter^3 g^{-3}. Migration from right to left.

of the theory. As illustrated in Fig. 33, the bimodal sedimentation and descending electrophoretic patterns of this protein under optimal conditions for tetramerization are strikingly similar. Tombs (40) has analyzed very highly purified β-lactoglobulin A electrophoretically over a wide range of protein concentration. Representative patterns are displayed in Fig. 36. As predicted theoretically, the absolute area of the slower descending peak remained constant, while that of the faster one increased with increasing total concentration. The bimodal reaction boundary is characterized by a Δ_s of 1.8 ± 0.2 g liter^{-1} as judged from the data points in Fig. 37. This value is to be compared with that of 1.5 ± 0.14 g liter^{-1} estimated by Townend and his co-workers (13) from sedimentation patterns. The agreement is quite satisfactory in view of the uncertainties in locating the minimum in bimodal reaction boundaries and the approximations inherent in the Gilbert theory.

While Gilbert's continuous solution of Eq. (197) applies to descending electrophoresis of systems for which $v_P > v_M$, for ascending electrophoresis of such systems, it describes a physically impossible situation (Fig. 38), corresponding to double-valued concentration at every position except where the front joins the plateau. It may even be that the continuous solution is mathematically unstable for conditions which obtain at the ascending boundary. In any case, Wilson's discontinuous solution of Eq. (197) is stable for these conditions. Thus, if the shape of the infinitely sharp initial boundary were to be perturbed during electrophoretic transport, polymer molecules in the dilute leading edge would dissociate into monomer molecules which lag behind, thereby resharpening the boundary. Now, Eq. (210) tells us that the discontinuous boundary is located at $\eta = \bar{v}^0$ or, in terms of our moving coordinate system, at

$$\delta_d = \frac{\bar{v}^0 - v_M}{v_P - v_M} = \frac{c_P^0}{c_M^0 + c_P^0} \tag{217}$$

where superscript zero specifies values in the plateau region behind the boundary. In other words, the infinitely sharp boundary migrates with the weight-average velocity of the material in the plateau, and Eq. (217) is to be interpreted to mean that the product $\delta_d v_P$, is the additional velocity imparted to the system by the presence of polymer. Thus, in the limit of infinite dilution, $c_P^0 = 0$, $\delta_d = 0$, and the boundary moves with velocity v_M, while, at infinitely high concentration, the material is completely polymerized, $\delta_d = 1$, and the boundary moves with velocity v_P. In the illustrative calculation presented in Fig. 38, the discontinuous boundary (vertical, broken line) is located at $\delta_d = 0.5$ because the material is polymerized to the extent of 50% by weight in the plateau.

It must be emphasized that the similarity solutions of Eq. (192) which we have been considering apply only to an infinitely sharp initial boundary. While this condition is realized in sedimentation, initial electrophoretic boundaries are generally blurred somewhat by diffusion. It is pertinent, therefore, to seek a solution for any initial conditions. Such a solution is

$$x = S(c_M) + t \frac{v_M + v_P n K' c_M^{n-1}}{1 + n K' c_M^{n-1}} \tag{218}$$

in which $S(c_M)$ defines the initial boundary [(23–25); see especially Thomas (25) for the solution of differential equations of this form]. The reader can easily verify that, for descending electrophoresis of systems characterized by $v_P > v_M$, this solution generates a Gilbert-type bimodal reaction boundary whose first moment is given by

$$\bar{x} = \bar{S}(c_M) + \bar{v}^0 t \tag{219}$$

where $\bar{S}(c_M)$ is the first moment of the initial boundary.[8] In other words, the

---

[8] Note that the particular solution for an infinitely sharp initial boundary is

$$x = t\lambda(c_M), \quad 0 < c_M < c_M^0$$
$$\infty \geq x \geq t\lambda(c_M^0), \quad c_M = c_M^0$$
$$v_M t \geq x \geq -\infty, \quad c_M = 0$$

where $\lambda(c_M) = (v_M + v_P n K' c_M^{n-1})/(1 + n K' c_M^{n-1})$. For sedimentation or descending electrophoresis of a system for which $v_P > v_M$, this solution describes a bimodal reaction boundary, $[\partial(c_M + c_P)/\partial x]_t$ vs. x, whose trailing edge is located at $v_M t$ and leading edge at $t\lambda(c_M^0)$, which is less than $v_P t$. The width of the boundary increases with time at the constant rate $(v_P - v_M)[(n K'(c_M^0)^{(n-1)})/(1 + n K'(c_M^0)^{(n-1)})]$, and its minimum is located at $x_{min} = t\{(v_P - v_M)[(n-2)/3(n-1)] + v_M\}$.

rate of migration of the first moment gives the weight-average velocity in the plateau region ahead of the boundary. This result brings unity to our overall discussion of electrophoresis. Note first of all that the relationship expressed by Eq. (219) was obtained previously from the weak-electrolyte moving-boundary theory presented in Chapter II [see Eqs. (98) and (109) and accompanying discussion]. Then, compare Eq. (219) with the first moment of the general solution [Eq. (184)] of the differential equation for the electrophoresis of an inherently homogeneous, noninteracting macromolecule under the assumption of negligible diffusion. The conclusion reached is that the first moment of any electrophoretic moving boundary, irrespective of shape, is the equivalent boundary position which conserves mass and which migrates with the constituent velocity of the corresponding component.[9] Longsworth (41) had already recognized this important principle a quarter of a century ago.

Now, when applying the solution given by Eq. (218) to ascending electrophoresis, the velocities are assigned negative values, since migration is from solution to solvent. When this is done, one finds that the initial boundary sharpens with time of electrophoresis until some critical time is reached after which the solution generates a physically impossible situation corresponding to many-valued concentration at positions in the front. The mathematical problem becomes an extremely difficult one (23–25) because there is no way of joining the continuous solution given by Eq. (218) to Wilson's discontinuous solution of Eq. (192). The difficulty stems from the fact that Wilson's discontinuous solution can be rigorously applied only when the boundary becomes infinitely sharp[10] [see the comments of Thomas (25), pp. 163 and 164 and particularly the one on p. 164 concerning the usual treatment of this problem in chromatography]. However, from physical considerations, it can

---

[9] The term "component" is defined here as the thermodynamic component for either an interacting system in which equilibrium is attained rapidly or a mixture of noninteracting macromolecules. For slow macromolecular reactions, the electrophoretic components are reactants and products. In the case of interactions characterized by rapid reequilibration, the constituent velocity is determined by the particular interaction between macromolecular species superimposed upon the whole set of acid–base and other small-ion-binding equilibria. For slow reactions and noninteracting mixtures, the constituent velocities are determined solely by the acid–base and ion-binding equilibria.

[10] It is instructive to examine the particular case of an initial ascending boundary identical to a Gilbert descending boundary generated by τ sec of electrophoresis. For this hypothetical situation, $S(c_M)$ is given by τ times the right-hand side of Eq. (203) with positive velocities. Now, we see from Eq. (218) (for which, velocities are negative in the second term) that, when the field is applied, the initial boundary sharpens continuously as it migrates along the column until it becomes infinitely sharp at time $t = \tau$. At that instant, the continuous solution joins rigorously to Wilson's discontinuous solution and the boundary remains infinitely sharp forever after.

be seen that the boundary must sharpen continuously during electrophoresis since, under the influence of mass action, polymer molecules in the dilute leading edges dissociate into monomer molecules, which lag behind. Once the boundary has become infinitely sharp, it must remain so as it migrates along the electrophoresis column with the weight-average velocity in the plateau behind. Moreover, this mechanism compensates for diffusion spreading of actual ascending reaction boundaries encountered in practice, thereby maintaining their hypersharpness once it has been established.

A beautiful example of the predicted behavior is afforded by the electrophoretic patterns of β-lactoglobulin A displayed in Fig. 36. Whereas the descending pattern shows a bimodal reaction boundary, the ascending boundary is hypersharp.

The above considerations are restricted, of course, to systems for which $v_P > v_M$. In the event that $v_P < v_M$, the situation is exactly reversed: now the descending boundary is hypersharp and the ascending boundary is bimodal. In either event, the described nonenantiography is diagnostic of interaction and usually distinguishable from the Dole nonideal electrophoretic effects. Thus, hypersharpening of ascending boundaries associated with the generation of pH and conductance gradients by the electrophoretic process *per se* generally tends to enhance rather than obliterate resolution of noninteracting macromolecules. Having made the initial observation, supporting evidence for reversible polymerization is then sought by: electrophoretic analyses at progressively increasing protein concentration, as was done in the case of β-lactoglobulin A; fractionation experiments; and independent physicochemical measurements, such as velocity and equilibrium ultracentrifugation and light scattering.

Finally, the considerations presented above and in footnote 10 predict an unambiguous electrophoretic test for interactions. Suppose that an electrophoretic experiment of duration τ sec yields electrophoretic patterns such as those displayed in Fig. 36. If the field is reversed at that time, the hypersharp ascending boundary will generate a bimodal reaction boundary as it now migrates back to its original starting position. After τ sec of reverse electrophoresis, the first moment of the newly generated bimodal boundary will be located at the original starting position. Simultaneously, the bimodal descending boundary will sharpen as it migrates back to its original starting position; after τ sec, the newly generated hypersharp boundary will be located at the original starting position. The predicted behavior[11] is in contradistinction to that expected of a simple mixture of noninteracting macromolecules. In the

---

[11] Since preparation of the manuscript this prediction has been verified experimentally with β-lactoglobulin A in 0.1 M acetate buffer, pH 4.7, at 0°C (T. D. Stamato and J. R. Cann, unpublished).

latter case, coalescence of resolved boundaries in either limb of the Tiselius cell upon reversal of the field should yield a single boundary which is unimodal and symmetrical, although somewhat broader than the original starting boundary due to diffusion. Moreover, if, for one reason or another, resolution does not occur in one of the limbs during forward electrophoresis, it will not occur when the field is reversed. While we have described the results of field reversal as predicted for association–dissociation reactions, the test is not limited to such interactions, and has been applied (*42*) to BSA in acetate buffer, pH 4. The electrophoretic patterns of BSA in this solvent are reaction boundaries (*42a*) arising from interaction of the protein with undissociated acetic acid and possessing a fine structure due to the superimposed N–F transition of Aoki and Foster (*43*). Upon reversal of the field, the reaction boundary in each limb of the cell generated a final boundary which was not a unimodal, symmetrical one, but which showed as many as three distinct, although poorly resolved, peaks. This result indicated that an equilibrium was being continuously readjusted as the boundaries returned to their initial positions, and constituted one of the first pieces of evidence for the reversible complexing of proteins with undissociated buffer acid, with a concomitant increase in mobility without significant change in frictional coefficient. This type of interaction will be dealt with more fully in the next chapter.

Molecular Sieve Chromatography. Although Gilbert originally formulated his theory with sedimentation and electrophoresis in mind, he recognized the analogy between these and other methods of mass transport, such as countercurrent distribution and chromatography. The analogous theories were subsequently elaborated by Bethune and Kegeles (*44*) for countercurrent distribution and by Ackers and Thompson (*45*) for molecular sieve chromatography (gel filtration on columns of Sephadex or other gel), thereby establishing the formal relationship among the four transport methods. Because of its ease of execution and its power as a means of fractionation, purification, and characterization of biologically important molecules in accordance with their size, a few words concerning molecular sieve chromatography are in order. Since a detailed account of the theory would be somewhat repetitive, we shall restrict ourselves to its principal predictions and applications to the detection and characterization of association–dissociation reactions.

The principles of molecular sieve chromatography can be summarized as follows. When dry particles of crosslinked dextran are equilibrated with solvent, they imbibe a large amount of water and swell to form a gel. While salt ions and other small molecules can generally diffuse freely into the water within the gel, large molecules are impeded from entering because of steric hindrance of the gel network and, if sufficiently large, they will not enter the gel at all. The upper limit of penetrability depends upon the degree of crosslinking of the gel, which, in turn, determines its molecular sieve characteristics. Thus, for example, while globular proteins of molecular weight less than about 50,000 can penetrate Sephadex G-25, those of higher molecular weight are excluded from the interior of the gel. In contrast, proteins of molecular weight as high as about 500,000 can penetrate the less highly crosslinked Sephadex G-200. Penetration is, of course, a graded effect, since the gel's labrinthine interior makes different groupings of regions unavailable to different size molecules (rigorously, characterized by their Stokes radii rather than molecular weight). Consequently, the extent to which molecules of progressively increasing size penetrate a given gel varies continuously from free access to exclusion. Now, consider a column of gel equilibrated with solvent. Let us introduce into the top of the column a small volume of solution containing a mixture of three substances having different molecular weights: a freely penetrating, low-molecular-weight substance A; a macromolecular substance B with intermediate tendency to penetrate the gel; and a macromolecular substance C of sufficiently large size so as to be excluded. If we allow instantaneous establishment of equilibrium, the three substances immediately distribute themselves between the interior of the gel and the exterior mobile phase in such a fashion that (1) the concentration of substance A inside the gel is the same as in the mobile phase; (2) the concentration of substance B is lower inside than outside; and (3) substance C is confined to the mobile phase. Let us assume for the sake of argument that, at equilibrium, the concentration of substance B within the gel is one-half of its concentration in the mobile phase. When a steady flow of solvent is introduced into the column, a zone of each substance is transported down the column at a rate dependent upon its distribution between the stationary and mobile phases. Accordingly, substance C emerges from the column first, since it is merely necessary to replace the mobile phase in the column with an equal volume (the void volume of the column) of fresh solvent in order to elute an excluded substance. Substance B then emerges when the volume of eluent becomes equal to the void volume of the column plus one-half the internal gel volume of the column. Perhaps this is seen most clearly by recalling that the quantity of substance B on the column is equal to its concentration in the mobile phase times the sum of the void volume and one-half the internal gel volume corresponding to the width of its zone at any instant. It follows that the volume of eluent required to move the zone through the entire length of the column must be the sum of the void volume of the column and one-half of its internal volume. Finally, substance A emerges when the volume of eluent becomes equal to the total volume of the column, i.e., the sum of the void volume of the column and its total internal gel volume. Thus, the three substances have been separated one from another by virtue of differences in their ability to penetrate the gel.

The phenomenological parameter of molecular sieve chromatography is the molecular sieve coefficient[12] σ, which characterizes the interaction between a given molecular species and the gel. The coefficient is defined as the ratio of the equilibrium concentration of that species within the gel to its concentration in the mobile phase. The usual way of performing molecular sieve chromatography is to apply to the column such a small volume of the solution of macromolecules to be fractionated that the separated components elute as discrete peaks, i.e., the effluent profiles of the separated components do not show a plateau region. In that event, the molecular sieve coefficient of each component is calculated as

$$\sigma = (V_e - V_0)/V_i \qquad (220)$$

where the elution volume V_e is the volume of eluent required to move the peak concentration of that component from the top to the bottom of the column, V_0 is the void volume of the column, and V_i is its internal gel volume. The void volume is the elution volume of a solute known to be completely excluded from the gel particles, e.g., blue dextran, and the internal gel volume is the difference between the total volume V_t and the void volume. The former is the elution volume of either tritiated water, which penetrates the gel to the same extent as ordinary water, or some small solute of known σ close to that of water, e.g., KCl, with $\sigma = 0.98$ in Sephadex G-25. Thus, we see that the converse of the statements made in the preceding paragraph is that the ability to separate molecules of different size by gel filtration depends formally upon the differences between their molecular sieve coefficients. The smaller the difference, the greater is the column length required to effect separation.

Mathematical Description. The above concepts can be formulated quantitatively as follows. First, we take note that the mathematical description of molecular-sieve chromatography usually makes the simplifying assumptions of negligible diffusion and instantaneous equilibration of solute between mobile and stationary phases. The latter is approximately realized in practice by eluting the material from the column at a sufficiently slow rate. Given these assumptions, the differential equation describing the gel filtration of a single solute is derived (*23, 45*) by considering a small cross-sectional layer of the column of thickness dx and containing a certain distribution of solute between the mobile and stationary phases. The concentration of solute in the mobile phase is designated as c, and the amount of material Q occluded in the gel is determined by the molecular sieve coefficient to be $Q = \beta\sigma c$, where β is the

---

[12] Usually referred to as the distribution coefficient K_d, but we shall retain the nomenclature of Ackers and Thompson (*45*), since it simplifies the mathematical notation.

internal gel volume per centimeter length of column. The difference between the concentration at the rear of the cross-sectional layer and at its front is $-(\partial c/\partial x)_v \, dx$. When a volume dV of eluent passes into the volume, an equal volume of solution will enter the layer under consideration at its rear boundary, and, simultaneously, a different portion of solution of the same volume will leave across the front boundary. The amount of solute carried into the layer by the solution entering it exceeds the amount leaving by $-(\partial c/\partial x)_v \, dx \, dV$. At the same time, the amount of solute in the mobile phase of the layer increases by $\alpha \, dx (\partial c/\partial V)_x \, dV$, where α is the void volume per centimeter of column length, and the amount of occluded solute increases by $dx(\partial Q/\partial V)_x \, dV = dx \, \beta\sigma(\partial c/\partial V)_x \, dV$. Conservation of mass requires that the net accumulation of material equal the sum of the increase in mobile and occluded solute:

$$-(\partial c/\partial x)_v \, dx \, dV = \alpha \, dx(\partial c/\partial V)_x \, dV + dx \, \beta\sigma(\partial c/\partial V)_x \, dV \quad (221)$$

By canceling dx and dV and rearranging, we obtain the differential equation of molecular-sieve chromatography:

$$(\partial c/\partial x)v + (\alpha + \beta\sigma)(\partial c/\partial V)_x = 0 \quad (222)$$

Under the coordinate transformation $y = x - [V/(\alpha + \beta\sigma)]$, this equation becomes $(\partial c/\partial V)_y = 0$, which tells us that the initial spatial distribution is preserved as the zone of material moves down the column. Now, the general solution of the differential equation is

$$x = S(c) + [V/(\alpha + \beta\sigma)] \quad (223)$$

where $x = S(c)$ at $V = 0$ defines the initial shape of the zone. Imagine that such a small volume of solution of the solute is introduced into the column that the solute may be considered as being present initially only at the origin. In that event, $S(c) = 0$. The elution volume V_e is then found by setting x in Eq. (223) equal to the length l of the column, noting that $\alpha l = V_0$ and $\beta l = V_i$, and rearranging to obtain

$$V_e = V_0 + \sigma V_i \quad (224)$$

This equation is the formal embodiment of our earlier qualitative description of the separation of a mixture of three substances with different tendencies to penetrate the gel. Thus, $\sigma = 1$ for the freely penetrating substance A, so that its elution volume is equal to $V_0 + V_i$, the total volume of the column; $\sigma = \frac{1}{2}$ for substance B of intermediate tendency to penetrate the gel, and thus its elution volume is $V_0 + \frac{1}{2}V_i$; and $\sigma = 0$ for the excluded substance C, so that its elution volume is V_0, the void volume of the column.

Clearly, effective fractionation of a mixture requires a sufficiently large difference between the elution volumes of its constituents. Consider, for example, a mixture of two macromolecules designated by subscripts 1 and 2. From Eq. (223) and the relationship $V_i = \beta l$, we find that the difference in the elution volumes of the two molecules is

$$V_1 - V_2 = (\sigma_1 - \sigma_2)\beta l \qquad (225)$$

Accordingly, the smaller the difference between the molecular sieve coefficients, the greater is the column length required to effect separation.

Finally, if a relatively large volume V of solution of a single solute is applied to the column, the elution volume of the infinitely sharp trailing edge of the zone is still given by Eq. (224), while that of the leading edge is smaller by the amount V.

Reversibly Associating Systems. Let us now consider the gel filtration of associating–dissociating systems. The molecular sieve coefficient of a nonassociating protein such as ovalbumin is independent of column length. However, the apparent coefficient of an associating-dissociating protein like β-lactoglobulin or hemoglobin is expected to increase with increasing column length (45). This can be seen as follows. Since the peak spreads as it moves down the column, the protein concentration continuously decreases, with concomitant displacement of the equilibrium in favor of molecular dissociation. As the average size of the protein decreases, gel penetration increases; consequently, the elution rate decreases. Such variation of apparent coefficient with column length should provide a useful qualitative means of detecting molecular dissociation, but the resulting elution profiles can not be interpreted quantitatively in terms of the degree of dissociation.

The desired quantitative information can be obtained by the following change in experimental design: Suppose that the volume V of solute solution applied to the column is sufficiently large that the elution profile is a relatively broad band rather than a peak. Characteristically, the solute concentration in the plateau region of the band will be the same as the initial concentration. Now, when the band moves down the column, its leading and trailing edges continually "feed" from the constant plateau concentration as they spread. Consequently, their first moments move with constant velocity, independent of column length, despite molecular dissociation. We note the analogies between the trailing edge and a sedimenting or descending electrophotretic boundary and between the leading edge and an ascending electrophoretic boundary. It is possible, under these circumstances, to elaborate a mathematical theory of molecular sieve chromatography of rapidly associating–dissociating systems of the type $nM \rightleftharpoons P$ in a manner analogous to the Gilbert theory of sedimentation and electrophoresis.

Ackers and Thompson (45) have found the following expressions relating macromolecular concentration to position in the trailing edge of the elution profile:

$$c_M + c_P = \left(\frac{1}{nK'}\frac{\phi}{1-\phi}\right)^{1/(n-1)}\left[1 + \frac{1}{n}\frac{\phi}{1-\phi}\right] \quad (226)$$

$$\phi = \frac{\sigma_M - v'}{\sigma_M - \sigma_P} \quad (227)$$

$$v' = \frac{V - V_0 - \mathbf{V}}{V_i}, \quad V > V_0 + \mathbf{V} \quad (228)$$

which is identical in form to Gilbert's equation for sedimentation and electrophoresis [Eq. (211)]. In the equation for chromatography, σ_M and σ_P are the molecular-sieve coefficients of monomer and polymer, respectively, if they were to exist alone, and v' is a normalizing coordinate transformation. (Note that Ackers and Thompson reckon the total volume of the column starting at the leading edge of the initial zone of material.) Differentiation of Eq. (226) with respect to v' gives, for the gradient of total solute,

$$\frac{d(c_M + c_P)}{dv'} = \frac{1}{(n-1)(\sigma_M - \sigma_P)}\left(\frac{1}{nK'}\right)^{1/(n-1)}\phi^{(2-n)/(n-1)}\left(\frac{1}{1-\phi}\right)^{(2n-1)/(n-1)} \quad (229)$$

Thus, a unimodal gradient curve is predicted for dimerization, $n = 2$, and a bimodal one for higher-order polymerization, $n \geq 3$. The effluent volume of the rear of the gradient curve corresponds to σ_M, while that of the front is greater than for σ_P. The minimum in the bimodal gradient for $n \geq 3$ occurs at the effluent volume

$$V_{\min} = V_0 + \left[\frac{2n-1}{3(n-1)}\sigma_M + \frac{n-2}{3(n-1)}\sigma_P\right]V_i + \mathbf{V} \quad (230)$$

By substituting for $V_{\min}$ from Eq. (228) and rearranging, one obtains

$$n = (3v'_{\min} - \sigma_M - 2\sigma_P)/(3v'_{\min} - 2\sigma_M - \sigma_P) \quad (231)$$

from which n may be estimated, at least for $n < 10$. Finally, Eqs. (226) and (229) predict virtually the same concentration dependence of gradient shape and elution rates as does the Gilbert theory for the analogous quantities in sedimentation and electrophoresis.

The normalizing coordinate transformation for the leading edge of the band is

$$v = (V - V_0)/V_i, \quad V > V_0 \quad (232)$$

The leading edge is theoretically hypersharp and located at $v = \bar{\sigma}^0$, where $\bar{\sigma}^0$ is the weight-average molecular sieve coefficient in the plateau. In practice, the leading edge spreads somewhat, but its centroidal volume $\bar{V}$ is given by

$$\bar{V} = V_0 + [\alpha \sigma_M + (1 - \alpha)\sigma_P] V_i \tag{233}$$

which is also the weight average of the elution volumes of monomer and polymer in the plateau:

$$\bar{V} = \alpha V_M + (1 - \alpha) V_P \tag{234}$$

Thus, the weight fraction of monomer is given by

$$\alpha = (\bar{V} - V_P)/(V_M - V_P) = (\bar{v} - \sigma_P)/(\sigma_M - \sigma_P) \tag{235}$$

in which $\bar{v}$ is the centroid position of the leading edge. If n is known from some independent physicochemical measurement, the equilibrium constant can be calculated from Eq. (235) and the relation

$$K' = (1 - \alpha)/\alpha^n [c_M{}^0 + c_P{}^0]^{(n-1)} \tag{236}$$

Determination of accurate values of σ_M and σ_P presents a problem, of course. However, in favorable cases, they may be obtained from experiments at such low solute concentration that the protein is completely in the monomeric form, and at sufficiently high concentration that it is completely polymerized.

The essential features of the theoretically predicted behavior have been verified experimentally by gel filtration of carboxyhemoglobin (45) and α-chymotrypsin (46) under conditions chosen to optimize macromolecular association–dissociation. Thus, carboxyhemoglobin dissociates reversibly into its constituent subunits in the concentration range 5–900 μg ml^{-1}, at neutral pH and moderate ionic strength; and α-chymotrypsin associates strongly in 0.01 M phosphate buffer, pH 7.9, at concentration of the order of 1–5 mg ml^{-1}. The association of α-chymotrypsin in this buffer evidently can be described (1, 2) as one in which monomer coexists primarily with hexamer, the concentration of intermediate-size polymers being sufficiently low that the sedimentation pattern is bimodal. Typical elution profiles for these two proteins on Sephadex and the corresponding gradient curves are presented in Fig. 39.

It is immediately apparent that both profiles show a relatively sharp leading edge and a diffuse trailing one, which is the reverse of the behavior of a non-associating protein like ovalbumin. This feature is accentuated in the gradient curves, which show the same type of nonenantiography as the electrophoretic patterns (Fig. 36) of associating proteins, for which the polymer has a greater mobility than the monomer. Moreover, as predicted theoretically, the trailing gradient curve is unimodal for carboxyhemoglobin ($n = 2$ with some reservations) and bimodal for α-chymotrypsin ($n > 2$).

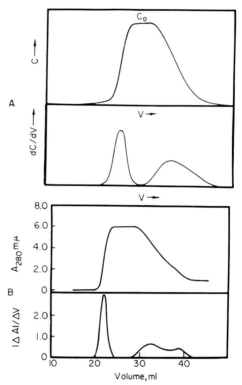

FIG. 39. Elution profiles and their derivative curves obtained in molecular sieve chromatography: (A) carboxyhemoglobin at a concentration of 5.0×10^{-6} mole liter^{-1} in 0.2 M sodium phosphate buffer, pH 6.8, on Sephadex G-25; (B) α-chymotrypsin, 3.8 mg ml^{-1}, in 0.01 M sodium phosphate buffer, pH 7.9, on Sephadex G-100. [Adapted from Ackers and Thompson (45) and Winzor and Scheraga (46).]

Ackers and Thompson (45) interpreted the leading edge of the elution profiles of carboxyhemoglobin quantitatively in terms of dissociation into half-molecules only. They obtained thereby a value for the dissociation constant in excellent agreement with that derived from integral diffusion data. There is, however, some ambiguity here. Thus, the photoelectric-scanner ultracentrifuge measurements of Schachman and Edelstein (47) suggest that the dissociation of human oxyhemoglobin under these conditions goes detectably beyond the half-molecule stage. Nor has the apparent discrepancy been resolved by the highly precise gel filtration experiments of Chiancone and co-workers (48) on oxyhemoglobin. In fact, the results of their sophisticated statistical analysis of the data illustrate the great difficulty of eliminating ambiguity from the calculation of characteristic parameters of reversibly reacting systems.

Quantitative interpretation of the elution profiles of α-chymotrypsin is subject to even greater uncertainty, since σ_P cannot be determined directly, and the stoichiometry of association is not sufficiently well understood to justify calculating σ_P from the position of the minimum in the trailing gradient curve. Nevertheless, useful information has been derived from these experiments. Consider, for example, the shape of the gradient curves of the trailing edge of elution profiles obtained at progressively increasing protein concentration. It is apparent from Fig. 40 that the shape of the gradient curves show the theoretically predicted concentration dependence. Thus, the single peak observed at low concentrations grows in area as the concentration is increased to some critical value corresponding to Δ_s in sedimentation and electrophoresis. Upon increasing the concentration still further, a second, more rapidly eluting peak appears and grows in area while that of the slower peak now remains constant. This same concentration dependence is expected of the velocity sedimentation patterns, but the diffuse nature of the α-chymotrypsin boundary under these conditions has precluded sedimentation experiments below a concentration of 3.5 mg ml^{-1} (*21*). Thus, comparison of gel filtration gradient curves with sedimentation patterns is possible only at the two highest concentrations, where quite good agreement is observed between the two methods. The concentration dependence of the rate of elution of the two peaks also conforms to theoretical expectations. While the rate of elution of the slower peak is predicted to be independent of concentration, that of the faster one should increase with increasing concentration. It is evident from Fig. 41a that this is an accurate description of the experimental findings. Turning our attention to the unimodal gradient curve corresponding to the leading edge of the elution profile, we recall that its mean rate of elution is theoretically the weight-average rate in the plateau region. That being the case, its elution rate should increase with increasing concentration. Once again, theory and experiment are in accord (Fig. 41b). For comparison, results for the nonassociating protein ovalbumin are also presented in this figure. As anticipated, very little variation of the elution rate was observed with ovalbumin. Clearly, such measurements provide a powerful means for detecting association–dissociation reactions, provided, of course, that the observations are carried to sufficiently low concentration (*49, 49a*).

The foregoing discussion has dealt almost exclusively with conventional procedures of molecular sieve chromatography, in which observations are made on integral boundaries between solution and solvent. Reference should be made, however, to the elegant differential methods of Gilbert (*50*). Here, attention is focused alternatively on: (1) the boundary formed in the gel column between two solutions of the same associating or otherwise interacting macromolecule at different concentrations, but under the same environmental conditions of pH, ionic strength, ligand concentration, and so on; (2) the dip

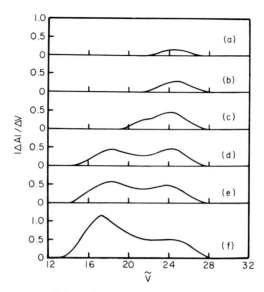

FIG. 40. Gradient curves of the trailing edge of the elution profiles of α-chymotrypsin at progressively increasing concentration. (a) 0.6 mg ml^{-1}; (b) 1.2 mg ml^{-1}; (c) 1.4 mg ml^{-1}; (d) 2.8 mg ml^{-1}; (e) 3.4 mg ml^{-1}; (f) 5.0 mg ml^{-1}. Other experimental conditions as in Fig. 39B. [From Winzor and Scheraga (46).]

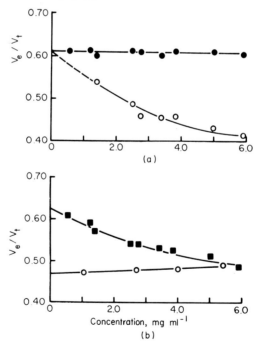

FIG. 41. The concentration dependence of the rate of elution of (a) the slower peak (●), and the faster one (○), in the trailing gradient curve of α-chymotrypsin; and (b) the leading gradient curve of α-chymotrypsin (■), and ovalbumin, (○). The rate of elution is inversely proportional to the ratio V_e/V_t. Experimental conditions as shown in Fig. 39B. [Adapted from Winzor and Scheraga (46).]

or hump localized in the elution profile at the junction between two solutions of the same protein at the same concentration, but subject to a difference in one of the environmental parameters; or (3) the dip or hump at the junction between a solution of a given protein, such as human hemoglobin, and another solution containing the same concentration of a closely related protein, like sheep hemoglobin, both in the same environment. In the first case, the elution volume of the differential boundary is a measure of the change in degree of association or dissociation caused by a change in concentration (*48*). In the other two cases (*50*), the appearance of the localized concentration perturbation implies, in the first instance, a difference in degree of dissociation caused by a difference in environmental parameter, and, in the second instance, an inherent difference in degree of dissociation. The sense of the perturbation indicates which solution is more highly dissociated, and the deficiency or excess of material which it represents is a measure of the difference in degree of dissociation. Since these methods lend themselves ideally to differential techniques of observation, they are capable of great sensitivity. Thus, they hold promise for the detection and characterization of small differences in degree of association–dissociation mediated, for example, by allosteric interaction with ligands (*50a*) or inherently related to genetic variation.

THE GILBERT–JENKINS THEORY

Reversible complex formation between different macromolecules is one of the most important of biological interactions, and the several methods of mass transport are particularly well suited for their detection and characterization. The application of electrophoresis to the study of the specific combination of pepsin with its macromolecular substrate, serum albumin, has already been described in some detail. Another example is the specific reaction of an antigen with its antibody. The significance of electrophoresis and ultracentrifugation for immunochemistry was recognized early. In fact, one of Tiselius' first applications of his moving-boundary electrophoresis apparatus was the demonstration (*51*) that rabbit antibodies to ovalbumin are associated with the γ-globulin fraction of the serum proteins. This pioneering experiment stimulated intensive programs for characterization of purified antibodies with respect to molecular parameters such as isoelectric point, electrophoretic mobility, molecular weight, and shape [see Isliker (*52*), Cann (*53*), and Kabat and Mayer (*54*) for summaries of these early experiments and more recent investigations], which, in turn, has led to important advances in the understanding of the nature of antibodies and the specific antigen–antibody reaction. The elegant studies of Singer and his co-workers (*55*) on the antigen–antibody reaction illustrates how moving-boundary electrophoresis and ultracentrifugation can help elucidate the nature of important biological reactions. Their experiments on soluble complexes formed between multivalent[13] protein antigens, Ag, and their specific antibodies, Ab, have established that (1) precipitating antibodies are bivalent, i.e., the antibody molecule possesses two sites, each of which can combine specifically with a complementary site on the antigen molecule; (2) the composition of the soluble antigen–antibody complex formed by reaction of antibody with a very large excess of antigen is Ag_2Ab; and (3) the framework theory of antigen–antibody precipitation is essentially correct. In addition, equilibrium constants and other thermodynamic parameters of the antigen–antibody reaction have been evaluated and a partial assessment of the forces responsible for the specific antigen–antibody bond has been possible.

Solutions containing only the protein antigen and its antibodies can be obtained by dissolving a specific antigen–antibody preicpitate in a sufficient excess of antigen. The antibody molecules in such solutions are not free, but are bound to antigen molecules in soluble antigen–antibody complexes of differing sizes. Although the various species are in equilibrium with one another and the rates of reaction evidently rapid, it has proven possible to interpret their electrophoretic and ultracentrifugal behavior in a more or less classical fashion without introducing excessive ambiguity. Ultracentrifugation is particularly well suited for studying specific complexing since it separates the various species primarily according to their size. In fact, Heidelberger and Pedersen (56) first demonstrated the existence of soluble antigen–antibody complexes in the inhibition, or antigen-excess, zone of the precipitin reaction by analyses of solutions of ovalbumin–rabbit antiovalbumin precipitates dissolved in excess ovalbumin.

Ultracentrifugation has also played a major role in the much more recent systematic studies of Singer and his co-workers, and has provided direct evidence for the essential correctness of the framework theory of antigen–antibody precipitation. Consider, for example, the sedimentation behavior of a solution containing bovine serum albumin and "normal" γ-globulin from an unimmunized animal at pH 7.5 and ionic strength 0.1. This solution will not contain serum albumin–γ-globulin complexes, since no specific antibodies are present and nonspecific aggregation does not occur under these conditions of pH and ionic strength. The velocity-sedimentation pattern of the solution will show two boundaries, the slower-sedimenting one corresponding to bovine serum albumin and the faster to "normal" γ-globulin, which has the higher molecular weight. Measurement of the area under the two boundaries permits determination of the relative proportions of albumin and "normal" γ-globulin in the solution. Now, let us consider the more complex situation

---

[13] The valence of an antigen or antibody refers to the number of specific antigen- or antibody-combining sites on the surface of the molecule.

presented by the solution obtained when one dissolves a bovine seruɯ albumin–antibovine serum albumin precipitate in excess bovine serum albumin. This solution contains uncombined serum albumin and various serum albumin–antiserum albumin complexes, but no free antibody. Even though reequilibration among the various species may be rapid, the sedimentation pattern should show several peaks, the slowest-sedimenting one corresponding to uncombined serum albumin and the more rapidly sedimenting ones being manifestations of specific complexes. Since the molecular weights of the complexes are greater than that of the antibody, their rates of sedimentation should also be greater than the rate characteristic of antibody. The sedimentation patterns shown in Figs. 42a–c establish the essential features of this picture. The slowest-sedimenting boundary in each of these patterns corresponds to bovine serum albumin, and no free antibody peak is observed. The faster-sedimenting peaks may be regarded as qualitative representations of either single specific antigen–antibody complexes or mixtures thereof. Finally, the patterns demonstrate that the size distribution of the antigen–antibody complexes depends upon the total concentration of antigen. Thus, in heavy antigen excess (Fig. 42c) one rapidly sedimenting peak, designated as a, predominates over all other fast peaks. In moderate antigen excess

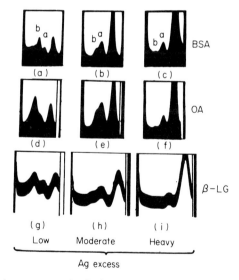

FIG. 42. Sedimentation patterns of soluble antigen–antibody complexes. Three patterns are shown for each of three different antigen–antibody systems; along a horizontal row, the degree of antigen excess decreases (the zone of precipitation is approached) from right to left. Within each pattern, sedimentation proceeds to the left. Phosphate buffer, pH 7.6, and ionic strength 0.1. The abbreviation BSA is for bovine serum albumin; OA, ovalbumin; and β-LG, β-lactoglobulin. [From Singer (57).]

(Fig. 42b), a number of faster-sedimenting peaks appear; in low antigen excess, close to the zone of precipitation, the faster peaks are most prominent (Fig. 42a). Although interpretation of the various rapidly moving peaks in terms of single antigen–antibody complexes of a given composition is open to some question, the a peak appears to correspond largely to a complex with the composition Ag_2Ab, at least in the regions of heavy antigen excess, and the b peak in Figs. 42a–c to Ag_3Ab_2. In any event, there can be no doubt that the growth of faster-moving peaks with decreasing antigen concentration in the region of antigen excess is due to the growth of larger antigen–antibody complexes. Strikingly similar results have been obtained with systems containing ovalbumin (Figs. 42d–f) or β-lactoglobulin (Figs. 42g–i) and their rabbit antibodies, which establishes the generality of these phenomena. It is clear that these results are a direct demonstration of the build up of a framework of alternating antigen and antibody molecules as the zone of precipitation is approached from the region of antigen excess. As the complexes grow beyond a certain size, they become insoluble, and a precipitate forms.

Systems of soluble antigen–antibody complexes have also been examined by moving-boundary electrophoresis. Electrophoretic analyses have yielded important information concerning the valence of precipitating antibodies, the thermodynamic properties of the protein antigen–antibody reaction, and the effect of pH on the combination of antigen and antibody. Some of these experiments have involved the concomitant use of electrophoretic and ultracentrifugal measurements, but we need only discuss the simplest electrophoretic experiments in any detail.

In contrast to ultracentrifugation, which separates macromolecules primarily according to their molecular weight, electrophoresis separates them primarily according to their net electrical charge. Let us consider the electrophoretic behavior of a solution containing bovine serum albumin and "normal" γ-globulin in barbital buffer at pH 8.5 and ionic strength 0.1. Since the serum albumin molecule carries a much larger negative net electrical charge than γ-globulin under these conditions, it will migrate much more rapidly toward the anode during electrophoresis than will γ-globulin. Consequently, the electrophoretic patterns of the solutions will show a fast-moving boundary corresponding to albumin and a slow-moving one corresponding to γ-globulin. Ideally, the area enclosed by each boundary is proportional to the concentration of the corresponding component. The situation with solutions obtained by dissolving a bovine serum albumin–antibovine serum albumin precipitate in excess serum albumin differs in that such solutions contain uncombined antigen and antigen–antibody complexes but no free antibody. Since the surface areas of the complexes are approximately the sum of the areas of the constituent molecules, the net electrical charge per unit area on the complexes are expected to be intermediate between those on serum albumin and antibody, and, therefore, their mobilities will be intermediate. Now, the ascending electrophoretic pattern should show a fast-moving boundary corresponding to uncombined serum albumin and one or more peaks which are manifestations of the specific complexes. The albumin boundary should migrate at the same rate as pure albumin in the absence of antibody, while the peaks which manifest the complexes should migrate at rates intermediate between those of the unbound albumin and pure antibody. As for the descending pattern, the fast-moving peak is a reaction boundary migrating with the constituent mobility of albumin in the reaction mixture, which will be slightly less than the mobility of uncombined albumin. In addition, there will be the slower peaks which manifest the complexes.

The observed electrophoretic patterns of several protein antigen–antibody systems are entirely consistent with this picture. Patterns shown by a bovine serum albumin–rabbit antibody system are presented in Fig. 43. Correction of the area of the fast-moving ascending boundary in these patterns for Dole's nonideal electrophoretic effects permits estimation of the equilibrium concentrations of uncombined antigen in solutions of known total antigen and total antibody content. Such data can be treated in much the same manner as described in Chapter II for the binding of small molecules by proteins in order to determine the number of specific binding sites on the antibody molecule. When this is done (Fig. 44), it is found that precipitating antibodies to bovine serum albumin, ovalbumin, and benzenearsonic acid are bivalent. This result agrees with the finding obtained by the entirely different method of equilibrium dialysis, that precipitating antibodies to simple haptens are bivalent. It also supports the conclusion from ultracentrifugal analyses that the principal antigen–antibody complex in heavy-antigen excess has the composition Ag_2Ab. *A priori*, one would expect the principal complex in heavy antigen excess to be one in which a single antibody molecule is bound to as many antigen molecules as its valence permits.

The thermodynamic properties of the protein antigen–antibody reaction can also be derived from these data. Although one cannot determine the equilibrium concentrations of a sufficient number of the various species coexisting in such solutions to permit direct evaluation of equilibrium constants, recourse is had to the Goldberg theory of the antigen–antibody reaction (59). Application of that part of the Goldberg theory which deals with systems in homogeneous equilibrium allows the indirect determination of the equilibrium constants from a knowledge of the equilibrium concentration of uncombined antigen, the total antigen and total antibody concentrations, and the valence of the antigen. By making reasonable assumptions concerning the latter, Singer and Campbell (60) have evaluated the equilibrium constant

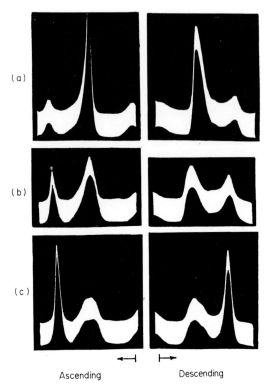

FIG. 43. Electrophoretic patterns of soluble antigen–antibody complexes at the following values of the weight ratio of total antigen to total antibody in solution, total protein concentration, and times of electrophoresis, respectively: (a) 0.56, 18.0 mg ml^{-1}, 7740 sec; (b) 1.00, 16.1 mg ml^{-1}, 7800 sec; and (c) 2.14, 16.5 mg ml^{-1}, 7560 sec. The antigen is bovine serum albumin. The initial boundaries are indicated by the vertical bar on the arrows. Barbital buffer, pH 8.6, and ionic strength 0.1. [From Singer and Campbell (58).]

for the reaction between uncombined antigen (bovine serum albumin or ovalbumin) and the complex AgAb to form the complex Ag_2Ab,

$$Ag + AgAb \leftrightarrows Ag_2Ab \tag{237}$$

at 0°C, in barbital buffer at pH 8.6 and ionic strength 0.1. (The values of the equilibrium constant computed for this reaction are essentially independent of the valence of the antigen, for a valence greater than 4.) Likewise, Pepe and Singer (61) have evaluated $K_i = K_1/2 = 2K_2$ for the only two possible reactions

$$Ag + Ab \leftrightarrows AgAb, \quad K_1 \tag{238}$$

$$Ag + AgAb \leftrightarrows Ag_2Ab, \quad K_2 = K_1/4 \tag{239}$$

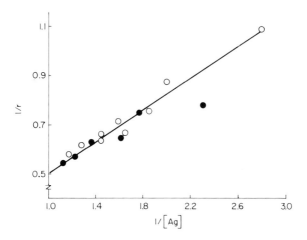

FIG. 44. Binding of protein antigens by antibody. Here, [Ag] represents the concentration of uncombined antigen divided by the concentration of total antigen; r is the molar ratio of antigen to antibody bound in all the complexes of a given solution. Extrapolation of the straight line to zero value of 1/[Ag] gives the reciprocal of the valence of the antibody. Open circles, BSA system; solid circles, OA system. [From Singer (57).]

between the univalent protein antigen, BSA–S–R_1, and purified anti-R antibody where BSA–SH represents bovine serum albumin;

$$R_1 = CH_2CONH-\langle\!\!\!\!\bigcirc\!\!\!\!\rangle-AsO_3H_2$$

and

$$R = \langle\!\!\!\!\bigcirc\!\!\!\!\rangle-AsO_3H_2$$

When evaluating these results, it must be borne in mind that the assumption has been made that the antibody-combining sites are homogeneous with respect to their intrinsic affinity for antigenic sites. Actually, however, the sites may be heterogeneous. If there is an essentially continuous distribution of intrinsic combining affinities among the sites, the value calculated by Singer and his co-workers for the equilibrium constant is roughly the average value for the whole population of antibody-combining sites. In that event, the calculated equilibrium constant should increase with decreasing antigen concentration, other things remaining constant. This is, in fact, the case (62).

Evaluation of the standard thermodynamic functions for the above reactions reveals that they are typically entropy driven. Thus, for example, the values of the several functions for the bovine serum albumin–antibovine serum albumin antibody system are: $\Delta F° = -5.5 \pm 0.2$ kcal mole^{-1}, $\Delta H° = 0 \pm 2$ kcal mole^{-1}, and $\Delta S° = 20 \pm 8$ cal deg^{-1} mole^{-1}. The small value of $\Delta F°$

shows that, in spite of their high specificity, antigen–antibody bonds are relatively weak. (The value of $\Delta F°$ for the reaction of H^+ and OH^- to form liquid H_2O is about -19 kcal mole^{-1}.) This finding is entirely consistent with the earlier conclusions of Pauling and his co-workers (63, 64) based on hapten inhibition of specific precipitation. They presented convincing evidence that the specificity of the antigen–antibody reaction can be explained in terms of the ordinary nonspecific, short-range forces that operate between all molecules, the specificity itself resulting from complementariness in structure of the antigen- and antibody-combining sites. That the specific bond is stabilized by weak forces such as electrostatic and van der Waals interactions and hydrogen bonds finds further support in the negligibly small $\Delta H°$. Thus, we see that the favorable free energy change is caused, in large part, by an increase in entropy. A priori, one might have predicted a decrease in entropy upon formation of antigen–antibody complexes because the number of particles in the system, and thus its randomness, would be expected to decrease. In fact, one might expect the loss in entropy upon association to be quite high because of the decrease in translational and rotational degrees of freedom which these molecules undergo. The expected change in entropy can be estimated from statistical-mechanical considerations to be about -120 eu, whereas the observed value is about $+20$ eu. This discrepancy is explicable in terms of the release of bound water from the antigen and antibody molecules upon formation of the specific bond. In that event, the reaction would be accompanied by an increase in the number of particles in the system, and thus in its randomness, and consequently the entropy would increase. If it is assumed that the entropy gained by the freed water is the same as that gained upon the analogous melting of ice, namely, 5 eu, the number of water molecules released is estimated to be about $\frac{1}{5}(120 + 20) = 28$. This represents a large release of water. Some of it is probably water of electrostriction released upon formation of a salt linkage between oppositely charged groups in the two combining sites. In fact, the influence of pH on the sedimentation and electrophoretic patterns of solutions of soluble complexes implicates a negatively charged carboxylate group in such a linkage. However, other mechanisms must also contribute. It has been suggested that the steric complementarity of antigen- and antibody-combining sites is associated not only with bringing a pair of oppositely charged groups into close proximity (when salt linkages are implicated in the specific bond), but also with obtaining the maximum release of bound water from the rest of the two combining sites, thereby assuring the maximum increase in entropy. These considerations ignore the possibility that some of the observed entropy gain might be due to configurational changes in the protein molecules, but physicochemical studies on antigen–antibody systems indicate that this certainly could not account for the major part of the total entropy change.

The Gilbert–Jenkins Theory

The foregoing not only attests to the power of ultracentrifugation and electrophoresis as aids in elucidating the mechanisms of specific interactions between biologically important macromolecules, but also points up the need for an appropriate transport theory to serve as an interpretative guide. It is particularly significant, therefore, that Gilbert and Jenkins (4) have solved the approximate transport equations for the interaction

$$A + B \rightleftharpoons C \tag{240}$$

where A and B are different macromolecules which react reversibly to form the complex C. In addition to describing the shape of the reaction boundaries encountered with such systems, their theory resolves some of the ambiguity attending the more-or-less classical interpretation of the sedimentation and electrophoretic patterns. Thus, for example, it becomes possible to estimate the error incurred when the area of the fast-moving BSA–S–R_1 boundary in the ascending electrophoretic pattern of BSA–S–R_1:anti-R complexes (Fig. 47b) is taken as a measure of the concentration of uncombined antigen in the original equilibrium mixture.

Formulation and Applications of the Theory. Like the Gilbert theory of mass transport of reversibly polymerizing systems, the analogous theory of Gilbert and Jenkins (4) for macromolecular-complex formation neglects diffusion. Accordingly, the transport-interconversion equations assume the form

$$\partial C_A / \partial t = -v_A (\partial C_A / \partial x) - R \tag{241}$$

$$\partial C_B / \partial t = -v_B (\partial C_B / \partial x) - R \tag{242}$$

$$\partial C_C / \partial t = -v_C (\partial C_C / \partial x) + R \tag{243}$$

The subscripts designate the reactants and their complex in the reaction (240), and R is the net rate of formation of the complex. The theory is elaborated for rates of reaction sufficiently large compared to the rate of separation of these species in the external field that local equilibrium obtains at every instant. Thus,

$$C_C = K C_A C_B \tag{244}$$

where K is the equilibrium constant for complex formation expressed in terms of molar concentrations. Elimination of R, C_C, and v_B from the above four equations yields the partial differential equation in the two dependent variables C_A and C_B,

$$KC_A \frac{\partial C_B}{\partial t} + (1 + KC_B) \frac{\partial C_A}{\partial t} = -(v_A + v_C K C_B) \frac{\partial C_A}{\partial x} - v_C K C_A \frac{\partial C_B}{\partial x} \tag{245}$$

Using the general theory of one-parameter groups of transformations, we find the similarity variables to be

$$\eta = x/t, \qquad C_A(\eta) = C_A(x, t), \qquad C_B(\eta) = C_B(x, t) \tag{246}$$

which transform Eq. (245) into the ordinary differential equation

$$[\eta(1 + KC_B) - (v_A + v_C KC_B)]\frac{dC_A}{d\eta} + KC_A(\eta - v_C)\frac{dC_B}{d\eta} = 0 \tag{247}$$

Multiplying through by $d\eta/dC_B$ gives

$$[\eta(1 + KC_B) - (v_A + v_C KC_B)]\frac{dC_A}{dC_B} + KC_A(\eta - v_C) = 0 \tag{248}$$

Let us now put $\mu = dC_A/dC_B$, which can be evaluated as follows: Applying the transformation given by Eq. (246) to Eqs. (241) and (242), we obtain

$$(\eta - v_A)\, dC_A/d\eta = Rt \tag{249}$$

$$(\eta - v_B)\, dC_B/d\eta = Rt \tag{250}$$

[Note that, whereas Eq. (246) is a similarity transformation for Eq. (245), this is not so for Eqs. (241) and (242), since their transformed equations involve t explicitly.] It follows immediately that

$$\mu = \frac{dC_A}{dC_B} = \frac{dC_A/d\eta}{dC_B/d\eta} = \frac{\eta - v_B}{\eta - v_A} \tag{251}$$

and therefore that

$$\eta = (\mu v_A - v_B)/(\mu - 1) \tag{252}$$

Substitution of μ for dC_A/dC_B and Eq. (252) for η in Eq. (248) yields the ordinary differential equation

$$(1 - \lambda)C_B\mu^2 + [(1 - \lambda)C_A + (1 + \lambda)C_B + 2/K]\mu + (1 + \lambda)C_A = 0 \tag{253}$$

where

$$\lambda = [2v_C - (v_A + v_B)]/(v_A - v_B) \tag{254}$$

This quadratic equation in μ has two solutions, but one of these can always be excluded on physical grounds. The value assigned to λ determines the form of the solution (applicable only to an infinitely sharp initial boundary) and hence the nature of the predicted concentration and concentration gradient curves. In accordance with the convention that velocities are positive in the direction solvent → solution and $v_A \geq v_B$, there are three main cases of interest:

Case (i): $v_C > v_A > v_B$, $\lambda > 1$
Case (ii): $v_A > v_C > v_B$, $-1 < \lambda < 1$
Case (iii): $v_A > v_B > v_C$, $\lambda < -1$

and also subsidiary cases such as $v_A = v_B \neq v_C$.

The Gilbert–Jenkins Theory

Here we shall concern ourselves only with the first two cases, which apply, respectively, to situations frequently encountered in the sedimentation and the electrophoresis of macromolecular complexes. Nor is there need to give explicit mathematical expressions for the similarity solutions of the differential equation. The reader is referred to the original paper of Gilbert and Jenkins (4) for a complete discussion of the mathematics and to the review of Nichol and his co-workers (28) for a summary along with appropriate experimental examples. It suffices here to examine a few of the sedimentation and electrophoretic patterns computed from these solutions and to make comparison with certain experimental patterns. From these considerations, there will emerge some guidelines for the interpretation of the sedimentation and electrophoretic behavior of reversibly complexing systems of macromolecules.

Let us first examine the gross features of the theoretical sedimentation patterns computed for the case $v_C > v_A > v_B$ and displayed in Fig. 45. We note

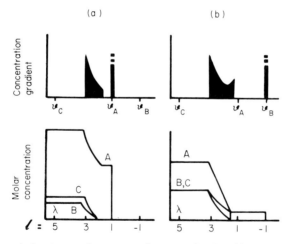

FIG. 45. Theoretical sedimentation patterns for a complex C and its components A and B, Case (i), $v_C > v_A > v_B$, $\lambda = 5$: (a) $KC_A{}^0 = 1.5$, $KC_B{}^0 = 0.25$; (b) $KC_A{}^0 = 1.0$, $KC_B{}^0 = 0.5$. Sedimentation proceeds from right to left. The abscissa is the dimensionless position variable $l = [\eta - \frac{1}{2}(v_A + v_B)]/\frac{1}{2}(v_A - v_B)$, on which scale $v_C \equiv \lambda$, $v_A \equiv +1$, and $v_B \equiv -1$. These patterns also apply to descending electrophoresis. [Adapted from Gilbert and Jenkins (4).]

first of all that, for a system characterized by a given equilibrium constant, the patterns show two boundaries over a wide range of composition of the reaction mixture. The slower-sedimenting boundary corresponds to pure reactant, either A or B, depending upon their relative equilibrium concentrations in the plateau region of the ultracentrifuge cell. This behavior is understandable in terms of mass action, for, as the complex sediments, leaving behind the two reactants, more complex is formed continuously until the least-concentrated reactant is exhausted. Accordingly, the slower boundary corresponds to the reactant initially present in excess. Likewise, the area of this boundary is less than would be expected from classical considerations, i.e., less than anticipated from the concentration of the reactant in the plateau. In practice, one generally characterizes an interacting system by varying the ratio of constituent concentrations of the reactants used to prepare the initial reaction mixture. If this ratio is varied over a sufficiently wide range, the relative equilibrium concentrations of the reactants will be markedly affected, and one can expect the boundary corresponding to one of the reactants to be replaced by one corresponding to the other. Although the particular reactant corresponding to the slower boundary may be identified through its sedimentation coefficient, it is desirable to confirm the identification by fractionation. For this purpose, the material disappearing across the slower-sedimenting boundary can be separated cleanly from faster-sedimenting material by means of the Yphantis–Waugh partition cell (67) and then analyzed.

Next, we note that the faster-sedimenting of the two boundaries shown by the theoretical patterns is a reaction boundary within which the equilibrium $A + B \rightleftharpoons C$ is continually readjusting during differential transport of reactants and complex. This boundary may itself be bimodal, as in Fig. 45b, and sediments at a velocity greater than those of the individual reactants, but considerably less than the velocity of the complex if it were to exist alone. Thus, we see that the individual peaks which constitute a reaction boundary may not be placed into correspondence with either reactant or complex, although the area of a given peak may sometimes prove to be an empirical index of the extent of reaction. This conclusion applies to electrophoresis as well as sedimentation.

The parametric conditions for which the theoretical patterns were computed might correspond, for example, to those which obtain in the case of a soluble complex formed by reaction of a relatively small antigen molecule with its antibody. While the simplest antigen–antibody systems known involve formation of two specific complexes, AgAb and Ag_2Ab, the reaction of the univalent antigen BSA–S–R_1 (molecular weight 70,000, 4 S) with anti-R antibody (molecular weight 150,000, 6 S) may be approximately treated as if only one complex were formed (68). This is possible because (1) light-scattering measurements indicate that the intrinsic association constant for this system is relatively small, and (2) the two complexes formed in accordance with reactions (238) and (239) have fairly similar sedimentation coefficients and may effectively be treated as one, to a first approximation. Accordingly, the sedimentation pattern of this system under the conditions of excess antigen (Fig. 46) may with confidence be compared with the theoretical pattern of Fig. 45b. The slower-sedimenting boundary corresponds to pure antigen and

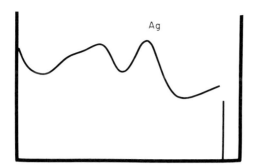

FIG. 46. Sedimentation pattern of a mixture of BSA–S–R_1 and anti-R antibody: 7.6 mg protein ml^{-1}; molar ratio of total BSA–S–R_1 to total anti-R is 2. Sedimentation proceeds to the left. [From Pepe and Singer (61).]

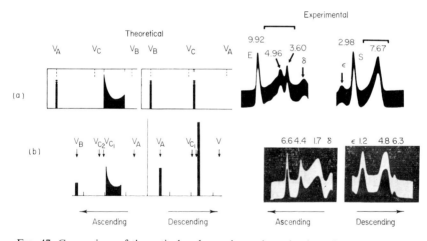

FIG. 47. Comparison of theoretical and experimental moving-boundary electrophoretic patterns for specific complex formation between (a) pepsin and bovine serum albumin, and (b) BSA–S–R_1 and anti-R antibody. Experimental conditions: (a) same as in Fig. 29, Chapter II; and (b) weight ratio of total BSA–S–R_1 to total anti-R_1, is 0.50; total protein concentration is 17.9 mg ml^{-1}; ionic strength, 0.1, barbital buffer, pH 8.70; 75 min of electrophoresis at 20.2 V cm^{-1}; mobilities of antigen and antibody about 6.6×10^{-5} and 1.1×10^{-5} cm^2 sec^{-1} V^{-1}, respectively. Brackets in pepsin–albumin patterns designate reaction boundaries. Theoretical patterns: (a) Case (ii) for a complex C and its components A and B, $v_A > v_C > v_B$, $\lambda = 0$, $KC_A{}^0 = 1.0$, $KC_B{}^0 = 6.6$; and (b) computed for the simultaneous reactions (238) and (239), $K_1 = 4K_2 = 1 \times 10^5$ liters mole^{-1}, $\mu_A = 1.2 \times 10^{-5}$, $\mu_B = 6.6 \times 10^{-5}$, $\mu_{C_1} = 4.1 \times 10^{-5}$, $\mu_{C_2} = 4.5 \times 10^{-5}$, antigen molecular weight is 70,000, antibody molecular weight is 150,000, concentrations correspond to experimental ones, and subscripts A, B, C_1, and C_2 refer to Ab, Ag, AgAb, and Ag_2Ab, respectively. [Experimental patterns from Cann and Klapper (65) and Pepe and Singer (61), and theoretical ones from Gilbert and Jenkins (4) and Gilbert and Gilbert (66).]

is so designated, while the faster two peaks undoubtedly constitute a bimodal reaction boundary. It is possible that this comparison is valid for all the patterns obtained by Pepe and Singer for this system [see Fig. 4 of Pepe and Singer (61)]. Moreover, it seems reasonable to intrepret the sedimentation behavior of the more complicated multivalent antigen–antibody systems in a similar fashion. Thus, the slower boundary in each of the patterns presented in Fig. 42 corresponds to uncombined antigen, and the faster peaks could conceivably constitute a reaction boundary.

As described previously, macromolecular complexes generally have electrophoretic mobilities intermediate in value between those of the reactants. This is so for antigen–antibody complexes and the Michaelis–Menten complex between pepsin and bovine serum albumin. It is of considerable interest, therefore, to consider the theoretical electrophoretic patterns computed for the case $v_A > v_C > v_B$, displayed in Fig. 47a. Both the ascending and descending patterns show two boundaries, one corresponding to a pure reactant, the other being a reaction boundary. In the descending pattern, the slower boundary corresponds to pure reactant B, and the faster one is a unimodal reaction

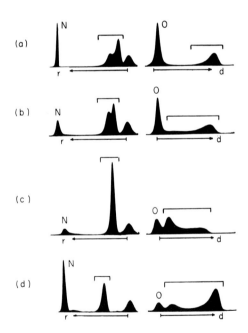

FIG. 48. Electrophoretic patterns of mixtures of ovomucoid (O) and yeast RNA (N) in 0.1 M sodium acetate buffers: (a) 0.87% O–0.55% N, pH 4.63; (b) 0.93% O–0.60% N, pH 4.33; (c) 0.93% O–0.60% N, pH 3.93; (d) 0.46% O–1.2% N, pH 3.92. Brackets designate reaction boundaries. [From Longsworth (69).]

boundary whose mobility is intermediate between those of the complex and reactant A. In contrast, the faster ascending boundary corresponds to pure A while the slower one is a bimodal reaction boundary with a mobility intermediate between the complex and reactant B. Although not evident from the figure, the areas of the faster-ascending and slower-descending boundaries are greater than expected from the equilibrium concentrations of the corresponding reactant in the plateau region behind or ahead of the companion reaction boundary [see Gilbert and Jenkins (*4*, Fig. 3) for plots of concentration versus position]. Such adjustment of concentration across a reaction boundary is a consequence of the reequilibration which accompanies differential transport of the various species. (This is not a unique feature of the Gilbert–Jenkins theory, however, for it is also predicted by the moving-boundary equation which is fundamental to the weak-electrolyte moving-boundary theory elaborated in Chapter II.) Finally, the similarities between the theoretical patterns and the experimental ones for the pepsin–serum albumin system are impressive. This is particularly so for the ascending bimodal reaction boundary. Other systems also show bimodal reaction boundaries. A striking example is the reaction between ovomucoid and yeast RNA in acidic media (Fig. 48). The patterns obtained at pH 4.63 and 4.33 are in

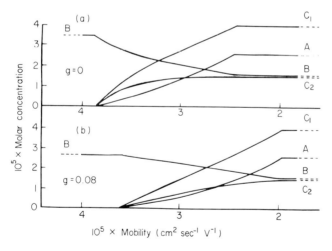

FIG. 49. Detailed analysis of the theoretical ascending electrophoretic pattern for the specific complexing of BSA–S–R_1 with anti-R antibody. Plot of the concentrations of the several species versus position expressed in mobility units. (a) Concentration curves corresponding to the ascending gradient pattern displayed in Fig. 46b, the mobility of each species assumed independent of the total concentration w of the solution through which it migrates; and (b) concentration curves computed for the situation in which the mobility of each species is concentration dependent according to the relation $\mu = \mu_0(1-gw)$, where $g = 0.08$ dl g^{-1}. Other symbols as in Fig. 46b. [From Gilbert and Gilbert (*66*).]

qualitative accord with the particular theoretical ones which we have just described. On the other hand, when the reaction occurs at pH 3.93, it is the descending, rather than the ascending, reaction boundary which is bimodal. This reversal in electrophoretic behavior must reflect an alteration in the parameters governing the interaction and the differential transport of the various species. In fact, the sharp pH dependence suggests an allosteric participation of hydrogen ions in the interaction.

Although the Gilbert–Jenkins theory is restricted to systems containing a single complex, it can be extended to include more complicated situations by resorting to numerical solution of the approximate transport equations. Gilbert and Gilbert (66) have made such calculations for the $BSA-S-R_1$: anti-R system in which only two antigen–antibody complexes are possible, reactions (238) and (239). Values of the various parameters were chosen so as to obtain approximate correspondence with some of the features of the electrophoretic data of Pepe and Singer (61). As shown by Fig. 47b, the agreement between the theoretical and experimental patterns is remarkable. Clearly, the experimental ones may be interpreted as follows. The faster-migrating ascending boundary corresponds to pure antigen, while the two slower peaks constitute a bimodal reaction boundary. The slower-descending boundary corresponds to pure antibody; and the major, more rapidly migrating boundary is a unimodal reaction boundary. It is important for the quantitative interpretation of the patterns to note that the area of the ascending antigen boundary is greater than expected from the equilibrium concentration of antigen in the plateau behind the companion reaction boundary (Fig. 49). It will be recalled that Singer and his co-workers took this area to be a classical measure of the equilibrium concentration of antigen in their determination of equilibrium constants. Singer et al. (68) subsequently showed, however, that the error thus incurred is not much greater than the experimental error, and that the apparent K values for the $BSA-S-R_1$: anti-R system calculated from electrophoretic data are close to the true value. This conclusion had, in fact, been reached independently (61) from the reasonably good agreement between equilibrium constants derived from electrophoresis and light-scattering measurements. They also concluded that, in multivalent antigen–antibody systems as well, the apparent equilibrium constants calculated from electrophoretic measurements are close in value to the true ones.

Finally, mention should be made of the adaptation of the Gilbert–Jenkins theory to the investigation of macromolecular complexes by the method of molecular sieve chromatography. Gel filtration has, of course, been widely used for investigating interactions of proteins with small molecules (70, 71), and it would seem to hold considerable promise as a means of characterizing interactions between macromolecules. In fact, the principles have already been enunciated and applied to a limited extent to the study of the simpler, rapidly established equilibria, namely, $nM \rightleftharpoons P$ (*45, 46, 48–50a*), $A + B \rightleftharpoons C$ (*72–78*), $A + B \rightleftharpoons C + D$ (*79*). But, as with other methods of mass transport, its potential will be fully realized only if used in combination with independent physicochemical measurements (*49*).

REFERENCES

1. G. A. Gilbert, *Discussions Faraday Soc.* **20**, 68 (1955).
2. G. A. Gilbert, *Proc. Roy. Soc.* (*London*) **A250**, 377 (1959).
3. G. A. Gilbert and R. C. Ll. Jenkins, *Nature* **177**, 853 (1956).
4. G. A. Gilbert and R. C. Ll. Jenkins, *Proc. Roy. Soc.* (*London*) **A253**, 420 (1959).
5. J. R. Cann, J. G. Kirkwood, and R. A. Brown, *Arch. Biochem. Biophys.* **72**, 37 (1957).
6. K. E. Van Holde, *J. Chem. Phys.* **37**, 1922 (1962).
7. W. F. Ames, "Nonlinear Partial Differential Equations in Engineering." Academic Press, New York, 1965.
8. J. Crank, "The Mathematics of Diffusion," pp. 148–149, 166–175. Oxford Univ. Press, London and New York, 1956.
9. I. B. Eriksson-Quensel and T. Svedberg, *Biol. Bull.* **71**, 498 (1936).
10. A. Tiselius and D. Gross, *Kolloid-Z.* **66**, 11 (1934).
11. A. G. Ogston and J. M. A. Tilley, *Biochem. J.* **59**, 644 (1955).
12. S. N. Timasheff and R. Townend, *J. Am. Chem. Soc.* **82**, 3157 (1960).
13. R. Townend, R. J. Winterbottom and S. N. Timasheff, *J. Am. Chem. Soc.* **82**, 3161 (1960).
14. R. Townend and S. N. Timasheff, *J. Am. Chem. Soc.* **82**, 3168 (1960).
15. R. Townend, L. Weinberger and S. N. Timasheff, *J. Am. Chem. Soc.* **82**, 3175 (1960).
16. S. N. Timasheff and R. Townend, *J. Am. Chem. Soc.* **83**, 464 (1961).
17. T. F. Kumosinski and S. N. Timasheff, *J. Am. Chem. Soc.* **88**, 5635 (1966).
18. J. J. Basch and S. N. Timasheff, *Arch. Biochem. Biophys.* **118**, 37 (1967).
19. P. Johnson and E. M. Shooter, *Biochim. Biophys. Acta* **5**, 361 (1950).
20. P. Johnson, E. M. Shooter and E. K. Rideal, *Biochim. Biophys. Acta* **5**, 376 (1950).
21. V. Massey, W. F. Harrington, and B. S. Hartley, *Discussions Faraday Soc.* **20**, 24 (1955).
22. J. N. Wilson, *J. Am. Chem. Soc.* **62**, 1583 (1940).
23. D. DeVault, *J. Am. Chem. Soc.* **65**, 532 (1943).
24. J. Weiss, *J. Chem. Soc.* **145**, 297 (1943).
25. H. C. Thomas, *Ann. N.Y. Acad. Sci.* **49**, 161 (1948).
26. L. W. Nichol and J. L. Bethune, *Nature* **198**, 880 (1963).
27. H. Fujita, "Mathematical Theory of Sedimentation Analysis." Academic Press, New York, 1962.
28. L. W. Nichol, J. L. Bethune, G. Kegeles, and E. L. Hess, in "The Proteins" (H. Neurath, ed.), Vol. II, 2nd ed., Chapter 9. Academic Press, New York, 1964.
29. A. G. Kirshner and C. Tanford, *Biochemistry* **3**, 291 (1964).
30. M. Fixman, *J. Chem. Phys.* **33**, 370 (1960).
31. G. A. Gilbert, *Nature* **186**, 882 (1960).
32. L. M. Gilbert and G. A. Gilbert, *Nature* **192**, 1181 (1961).
33. L. M. Gilbert and G. A. Gilbert, *Nature* **194**, 1173 (1962).
34. E. L. Smith, J. R. Kimmel, and D. M. Brown, *J. Biol. Chem.* **207**, 533 (1954).

35. W. L. Hughes, Jr., *Symp. Quant. Biol. Cold Spring Harbor* **14,** 79 (1950).
36. G. A. Gilbert, *Proc. Roy. Soc. (London)* **A276,** 354 (1963).
37. M. S. N. Rao and G. Kegeles, *J. Am. Chem. Soc.* **80,** 5724 (1958).
37a. J. L. Bethune and P. J. Grillo, *Biochemistry*, **6,** 796 (1967).
38. R. A. Brown and S. N. Timasheff, *in* "Electrophoresis Theory, Methods and Practice" (M. Bier, ed.), Chapter 8. Academic Press, New York, 1959.
39. J. R. Cann and R. A. Phelps, *Arch. Biochem. Biophys.* **52,** 48 (1954).
40. M. P. Tombs, *Biochem. J.* **67,** 517 (1957).
41. L. G. Longsworth, *J. Am. Chem. Soc.* **65,** 1755 (1943).
42. R. A. Phelps and J. R. Cann, *J. Am. Chem. Soc.* **78,** 3539 (1956).
42a. J. R. Cann, *J. Biol. Chem.* **235,** 2810 (1960).
43. K. Aoki and J. F. Foster, *J. Am. Chem. Soc.* **79,** 3385, 3393 (1957).
44. J. L. Bethune and G. Kegeles, *J. Phys. Chem.* **65,** 433, 1755, 1761 (1961).
45. G. K. Ackers and T. E. Thompson, *Proc. Natl. Acad. Sci.* **53,** 342 (1965).
46. D. J. Winzor and H. A. Scheraga, *Biochemistry* **2,** 1263 (1963).
47. H. K. Schachman and S. J. Edelstein, *Biochemistry* **5,** 2681 (1966).
48. E. Chiancone, L. M. Gilbert, G. A. Gilbert, and G. L. Kellett, *J. Biol. Chem.* **243,** 1212 (1968).
49. D. J. Winzor and H. A. Scheraga, *J. Phys. Chem.* **68,** 338 (1964).
49a. D. J. Winzor, J. P. Loke, and L. W. Nichol, *J. Phys. Chem.* **71,** 4492 (1967).
50. G. A. Gilbert, *Nature* **212,** 296 (1966).
50a. L. M. Gilbert and G. A. Gilbert, *in* "The Regulation of Enzyme Activity and Allosteric Interactions," pp. 73–87. Universitetsforlaget, Oslo, 1968.
51. A. Tiselius, *Biochem. J.* **31,** 1464 (1937).
52. H. C. Isliker, *Advan. Protein Chem.* **12,** 387 (1957).
53. J. R. Cann, *in* "Immunity and Virus Infection" (V. A. Najjar, ed.), p. 100. Wiley, New York, 1959.
54. E. A. Kabat and M. N. Mayer, "Experimental Immunochemistry," 2nd ed., Chapter 7. Thomas, Springfield, Illinois, 1961.
55. S. J. Singer, *in* "The Proteins" (H. Neurath, ed.), Vol. III, 2nd ed., Chapter 15. Academic Press, New York, 1965.
56. M. Heidelberger and K. O. Pedersen, *J. Exp. Med.* **65,** 393 (1937).
57. S. J. Singer, *J. Cell Comp. Physiol.* **50,** Suppl. 1, 51 (1957).
58. S. J. Singer and D. H. Campbell, *J. Am. Chem. Soc.* **74,** 1794 (1952).
59. R. J. Goldberg, *J. Am. Chem. Soc.* **74,** 5715 (1952).
60. S. J. Singer and D. H. Campbell, *J. Am. Chem. Soc.* **77,** 3499, 4851 (1955).
61. F. A. Pepe and S. J. Singer, *J. Am. Chem. Soc.* **81,** 3878 (1959).
62. B. W. Hudson, *Immunochemistry* **5,** 87 (1968).
63. L. Pauling, *Endeavour* **7,** No. 26 (1948).
64. D. Pressman and L. Pauling, *J. Am. Chem. Soc.* **71,** 2893 (1949).
65. J. R. Cann and J. A. Klapper, Jr., *J. Biol. Chem.* **236,** 2446 (1961).
66. L. M. Gilbert and G. A. Gilbert, *Biochem. J.* **97,** 7c (1965).
67. D. A. Yphantis and D. F. Waugh, *J. Phys. Chem.* **60,** 630 (1956).
68. S. J. Singer, F. A. Pepe, and D. Ilten, *J. Am. Chem. Soc.* **81,** 3887 (1959).
69. L. G. Longsworth, *in* "Electrophoresis Theory, Methods and Applications" (M. Bier, ed.), p. 91. Academic Press, New York, 1959.
70. P. Andrews, *Brit. Med. Bull.* **22,** 109 (1966).
71. G. F. Fairclough, Jr. and J. S. Fruton, *Biochemistry* **5,** 673 (1966).
72. L. W. Nichol and D. J. Winzor, *J. Phys. Chem.* **68,** 2455 (1964).

References

73. D. J. Winzor and L. W. Nichol, *Biochim. Biophys. Acta* **104,** 1 (1965).
74. L. W. Nichol and D. J. Winzor, *Biochim. Biophys. Acta* **94,** 591 (1965).
75. L. W. Nichol and A. G. Ogston, *Proc. Roy. Soc. (London)* **163B,** 343 (1965).
76. G. A. Gilbert, *Nature* **210,** 299 (1966).
77. D. J. Winzor, *Arch. Biochem. Biophys.* **113,** 421 (1966).
78. L. W. Nichol, A. G. Ogston, and D. J. Winzor, *J. Phys. Chem.* **71,** 726 (1967).
79. L. W. Nichol and A. G. Ogston, *Proc. Roy. Soc. (London)* **167B,** 164 (1967).

IV/ NUMERICAL SOLUTION OF EXACT CONSERVATION EQUATIONS

It is quite clear from the preceding discussion that the theoretical treatments of Gilbert and of Gilbert and Jenkins have had a profound influence on our concepts concerning the sedimentation and electrophoretic behavior of interacting systems, particularly with respect to the nature of reaction boundaries. In addition to providing the insight required for unambiguous analysis of the molecular mechanisms of certain types of interactions, these theoretical advances suggested the application of molecular sieve chromatography to such timely problems as the dissociation of hemoglobin into its subunits. And, like most advances of this sort, both theories pose new and interesting questions, some of which are concerned, for example, with allosteric interactions. This, in turn, has stimulated further theoretical investigations. By and large, these have made use of numerical methods to solve transport equations which are exact in the sense that the second-order diffusional term is retained. The mathematical methods are described in detail in Chapter V. Here, we shall consider their application to a variety of macromolecular interactions ranging in complexity from the relatively simple case of rapid dimerization to cooperative macromolecule–small-molecule interactions.

FACTORS GOVERNING THE PRECISE SHAPE OF REACTION BOUNDARIES

Without doubt, the most fundamental contribution of the analytic treatment of transport problems has been the demonstration that reaction boundaries may be bimodal even for rapidly established equilibria. It remains to define the effect of the various molecular and environmental parameters which may influence the precise shape of such boundaries. These include translational diffusion, the rates of reaction, in the case of kinetically controlled

Factors Governing the Precise Shape of Reaction Boundaries   153

processes, and the high pressures generated in the ultracentrifuge cell when the instrument is operated at moderate to high rotor speeds.

One's first concern naturally centers about the assumption of negligible diffusion, which must necessarily be made if the equations of transport are to be solved analytically as in the case of the Gilbert theory. In order to evaluate the influence of diffusion on the shape of reaction boundaries, one must resort to numerical integration of the exact differential equations of transport. The first efforts in this direction were evidently those of Field and Ogston (1), who analyzed the sedimentation behavior of hemoglobin under dissociating conditions. Satisfactory agreement was found between computed and experimental rates of spreading of the unimodal reaction boundary with time. At first glance, the rate of spreading is surprisingly insensitive to the dissociation–association reaction, the value of the apparent diffusion coefficient being only about 50% greater than expected from diffusion alone. But this result is consistent with the Gilbert theory, as judged from the expression given in footnote 8 of Chapter III for the rate of spreading of the boundary due to the reaction itself. The comparison reveals, however, one of the consequences of uncoupling diffusional and electrophoretic transport of reacting systems. While the Gilbert theory leads one to expect the apparent diffusion coefficient to increase with time of sedimentation, the numerical solution of Field and Ogston predicts the experimental observation that the apparent diffusion coefficient is independent of time. One hesitates to generalize this result, however, since in more complicated systems, the apparent diffusion coefficient may be time dependent (2) for one reason or another; further work in this direction would be desirable.

An important extension of the above approach to include higher-order association reactions was made by Bethune and Kegeles (3). As we shall see shortly, their computations demonstrate that Gilbert's assumption of negligible diffusion has only quantitative implications for the shape of the theoretical sedimentation patterns when monomer coexists in equilibrium with a single polymer, at least in the case of a trimerization reaction. In contrast, when monomer, dimer, and trimer coexist at comparable concentrations, the assumption appears not to be applicable. This investigation also illustrates how such computations aid in the determination of the stoichiometry of association reactions.

Bethune and Kegeles were concerned with the reversible association of α-chymotrypsin in 0.2 ionic strength phosphate buffer at pH 6.2, under which conditions, the sedimentation-velocity pattern shows a unimodal and symmetrical boundary (Fig. 50). Previously (4), the weight-average molecular weight of the protein in this solvent had been determined as a function of its concentration using the Archibald method. Taken alone, these latter data are interpretable within experimental error by assuming either monomer

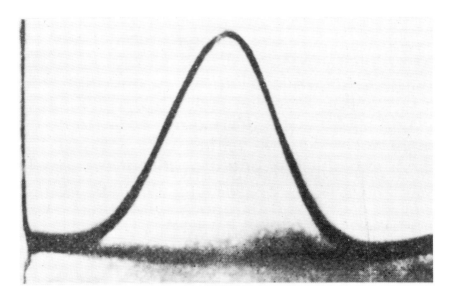

FIG. 50. Sedimentation-velocity pattern of α-chrymotrypsin in 0.2 ionic strength phosphate buffer, pH 6.2. Protein concentration, 18.9 g liter^{-1}; centrifugation for 120 min at 59,780 rpm; sedimentation proceeds from left to right. [From Bethune and Kegeles (*3*).]

coexisting with trimer alone, or the simultaneous presence of monomer, dimer, and trimer; equilibrium constants were calculated for both models. Given this information, comparison of theoretical and experimental sedimentation velocity patterns should permit a decision as to the actual stoichiometry of the association. In making this comparison, Bethune and Kegeles computed theoretical patterns for each model using the above-mentioned equilibrium constants. Moreover, comparison was made between patterns given by the Gilbert theory or a closed-form extension (*4*) thereof to the progressive polymerization reaction monomer ⇌ dimer ⇌ trimer, and patterns obtained by numerical solution of the exact conservation equations.

The analytic calculations are straightforward and require no further comment, but a few words concerning the numerical solutions are in order. The transport-interconversion equations for the system $3M \rightleftharpoons P$ are

$$\frac{\partial c_M}{\partial t} = D_M \frac{\partial^2 c_M}{\partial x^2} - v_M \frac{\partial c_M}{\partial x} + R_M \qquad (255)$$

$$\frac{\partial c_P}{\partial t} = D_P \frac{\partial^2 c_P}{\partial x^2} - v_P \frac{\partial c_P}{\partial x} - R_M \qquad (256)$$

It will be noted that, in contrast to Gilbert's treatment of this problem, the diffusion term has been included, although the equations still make the rectilinear approximation and assume constant velocities. As before, concentrations are expressed on a weight basis for convenience. Addition of the two equations gives the conservation equation

$$\frac{\partial(c_M + c_P)}{\partial t} = D_M \frac{\partial^2 c_M}{\partial x^2} - v_M \frac{\partial c_M}{\partial x} + D_P \frac{\partial^2 c_P}{\partial x^2} - v_P \frac{\partial c_P}{\partial x} \qquad (257)$$

Since we assume instantaneous reestablishment of equilibrium during differential transport, the effect of the association–dissociation reaction on the concentrations of monomer and polymer is given by

$$K_3' = c_P/c_M^3 \qquad (258)$$

Numerical solution of the latter two simultaneous equations was obtained on a high-speed electronic computer using a method which takes cognizance of the formal analogy between the finite-difference approximation to Eq. (257) and the expression for the first differences between the solute contents of adjacent tubes in a countercurrent distribution apparatus operating in a frontal mode. Essentially the same procedures were used for the progressive trimerization

$$2 \text{ monomer} \leftrightharpoons \text{dimer}$$

$$\text{dimer} + \text{monomer} \leftrightharpoons \text{trimer}$$

in which case, the conservation equation includes terms for the dimer, and the effect of the association reactions is given by two expressions of mass action.

The sedimentation patterns given by the analytical and numerical treatments are compared in Fig. 51. It is apparent that, for the model in which monomer coexists with trimer alone, inclusion of diffusional terms in the conservation equation has only a quantitative influence in smoothing out the bimodal reaction boundary. This result inspires confidence in the Gilbert theory of such systems. However, the comparison is not as favorable in the case of progressive polymerization. Whereas the extented Gilbert theory predicts a bimodal reaction boundary comprised of a major, rapidly sedimenting, asymmetrical peak and a small, asymmetrical slower one, the pattern given by the numerical solution shows a single peak with a small shoulder on its trailing edge. Moreover, the boundary might become symmetrical were account to be taken of the concentration dependence of the sedimentation coefficients. This is expected because monomer molecules in the dilute trailing edge would be speeded up while polymer molecules in regions of higher concentration would be retarded. In any case, the experimental pattern of α-chymotrypsin presented in Fig. 50 shows only a single peak; in view of the

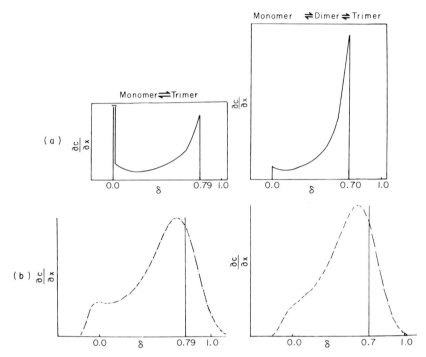

FIG. 51. Theoretical sedimentation velocity patterns computed for two different models of association of α-chymotrypsin under the conditions given in Fig. 50: (a) Gilbert theory and its extension to progressive trimerization; (b) transport theory which takes diffusion into account. For monomer in equilibrium with trimer alone, $c_M = 8.13$ g liter^{-1} and $c_T = 10.8$ g liter^{-1}; for simultaneous presence of monomer, dimer, and trimer, $c_M = 7.16$ g liter^{-1}, $c_D = 4.50$ g liter^{-1}, and $c_T = 7.24$ g liter^{-1}, where subscripts M, D, and T designate monomer, dimer, and trimer, respectively. [From Bethune and Kegeles (3).]

theoretical findings, the conclusion seems justified that comparable concentrations of monomer, dimer, and trimer coexist at equilibrium.

While the assumption of negligible diffusion is of little consequence for the major predictions of the original Gilbert theory (and probably also for the Gilbert–Jenkins theory), it is inadvisable to use this approximation in the treatment of more complicated types of association–dissociation reactions. This is so even for kinetically controlled processes, which one might expect *a priori* to be among the least sensitive of interactions. Consider, for example, the kinetically controlled dimerization $2M \underset{k_2}{\overset{k_1}{\rightleftharpoons}} P$, where k_1 and k_2 are the specific rates of association and dissociation, respectively. A unique feature of this system is that the theoretical sedimentation pattern may show three peaks whether or not diffusion is neglected. However, when diffusion is taken into account (5), the distinctive features of this system, found by solution of the approximate transport equations (6), are significantly altered. For example, alterations are found in the functional dependence of peak height and area upon time as well as in the range of relevant parameters over which three boundaries are observed. In particular, changes in total macromolecular concentration can induce remarkable changes in the patterns. As in the case of rapidly reversible polymerization and complex formation, this provides perhaps the most sensitive diagnostic test for interaction. The value of this test is further enhanced if the distribution of individual species can be detected, since either monomer or dimer can be bimodally distributed through the reaction boundary even though it may itself show one, two, or three peaks, depending upon the rates of reaction. When the half-times of reaction are small, the pattern corresponds to that of a dimerizing system for which local equilibrium obtains at every instant, and, accordingly, shows a single peak. For half-times of the order of the time of sedimentation, the pattern may show three peaks. Finally, if the rates of reaction are so small that significant re-equilibration does not occur during the time course of separation of the monomer and dimer, the pattern shows two peaks aping a mixture of two noninteracting macromolecules.

These results are reminiscent of those obtained previously for macromolecular isomerization, $A \underset{k_2}{\overset{k_1}{\rightleftharpoons}} B$. The two limiting cases corresponding to very rapid or very slow rates of interconversion were discussed early in Chapter II; the approximate analytical treatments for arbitrary rates of reaction were touched upon in the introduction to Chapter III. Here, we are concerned with accurate numerical solution of the exact transport-interconversion equations (82a, b) of Chapter II. A number of investigators (5, 7–12) have addressed themselves to this problem, since, in some respects, it represents the simplest of macromolecular interactions. More importantly, it provides insight into analogous situations posed, for example, by allosteric interactions of the type $A + nX \rightleftharpoons BX$ when the concentration of the small effector molecule X is virtually constant throughout the electrophoresis, sedimentation, or chromatographic column. The isomerization problem has been solved by several independent methods, e.g., numerical solution of the Fourier integrals which give the concentrations of the isomers (see the introduction to Chapter III) and solution of the finite-difference equations corresponding to the partial differential equations which describe the system. It is satisfying that they all predict the same behavior. In particular, the gradient curve can show one, two, or three peaks which constitute a single reaction boundary. Some representative computations illustrating the time course of development of the electrophoretic patterns are displayed in Figs. 52 and 53. The most striking feature of these results is the way in which the shape of the reaction boundary

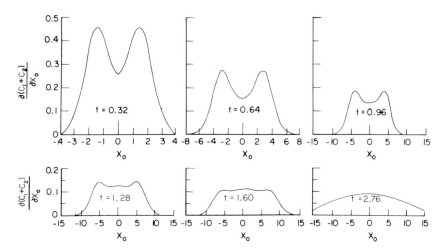

FIG. 52. Time course of development of theoretical moving-boundary electrophoretic patterns for an isomerizing system in which $D_1 = D_2$, $k_1 = k_2$, and $\alpha = 0.01$. Here, $\alpha = DK_1/(\mu_1 - \mu_2)^2 E^2$, $t_0 = k_1 t$, and x_0 is a reduced position variable. Ascending and descending patterns are enantiographic. [From Cann and Bailey (7).] The same predictions apply to chromatography and zone electrophoresis (8–11).

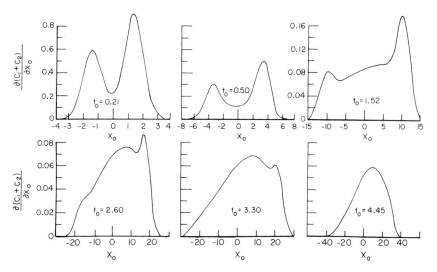

FIG. 53. Time course of development of theoretical moving-boundary electrophoretic patterns for an isomerizing system in which $D_1 = D_2$, $k_2/k_1 = \frac{2}{3}$, and $\alpha = 0.0049$. For $k_2/k_1 \neq 1$, the ascending and descending patterns are slightly nonenantiographic, but the essential features are the same in both. [From Cann and Bailey (7).] The same predictions apply to chromatography and zone electrophoresis (8–11).

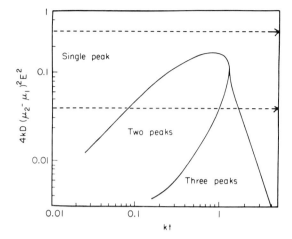

FIG. 54. Contour map defining the conditions for which the electrophoretic patterns of an isomerizing system will show one, two, or three peaks when $D_1 = D_2 = D$ and $k_1 = k_2 = k$. [Adapted from Scholten (12).]

changes with time of electrophoresis. Thus, for the particular values of the parameters used in these calculations, the boundary shows two peaks during early stages of electrophoresis. As electrophoresis proceeds, a third peak of intermediate velocity appears and grows continuously at the expense of the other two, until, eventually, the boundary becomes unimodal. The contour map presented in Fig. 54 defines the conditions for which the reaction boundary will show different numbers of peaks for the case in which the two isomers have the same diffusion coefficient and the rates of interconversion are the same in both directions. The horizontal broken lines are drawn to illustrate how the map is to be read. For the particular combination of values of the several molecular and kinetic parameters corresponding to the upper line, the pattern will show only a single peak at all times of electrophoresis because (1) the rates of reaction are so large that local equilibrium obtains at every instant, (2) the electrophoretic mobilities of the isomers do not differ significantly, or (3) diffusion obscures resolution. The lower line represents the combination of parameters used to compute the patterns displayed in Fig. 52. Finally, if the rates of interconversion are very small, large values of kt cannot be realized in practice. Consequently, the patterns will show two peaks at all times of electrophoresis. In this limit, the system behaves like a mixture of two noninteracting macromolecules. Although the foregoing description of isomerization has been presented in the context of moving-boundary electrophoresis, the conclusions reached are equally valid for zone electrophoresis (8, 9) and chromatography (8–11). They may also be applied to velocity and band sedimentation, to the rectilinear approximation, and, at least qualitatively, to zone velocity sedimentation through a preformed density gradient, provided the rates of interconversion are insensitive to the pressure generated in the ultracentrifuge cell at high rotor speeds. Indeed, the analogous proviso of pressure-insensitive equilibrium constant, while not stated previously, also applies to the Gilbert and Gilbert–Jenkins theories. When the ultracentrifuge is operated at moderate to high speeds, pressures of the order of 100–500 atm may be generated at the bottom of the cell. Consequently, if there is a change in volume upon reaction, the equilibrium constant will vary along the centrifuge cell according to the reaction isotherm, Eq. (79) in Chapter I. Since the pressure gradient in the cell is given by the relation

$$dP/dr = \rho \omega^2 r \tag{259}$$

the integrated form of the reaction isotherm is

$$\ln K(r) = \ln K(r_m) - \frac{\omega^2(r^2 - r_m^2)\rho}{2RT} \Delta \bar{V} \tag{260}$$

in which $K(r_m)$ is the equilibrium constant at the meniscus, where the pressure is only slightly greater than 1 atm. Thus, we see that the variation of the equilibrium constant from the meniscus to the bottom of the cell is determined by the molar volume change of reaction. The illustrative calculations presented in Table III show that this effect can be enormous for sufficiently large volume changes.

TABLE III

Effect of Pressure on Association Constant of Dimer and Biomolecular Complexes ($2A \rightleftharpoons A_2$ or $A + B \rightleftharpoons C$)[a]

Molecular weight of A [b]	Density of A	Density of B	Density of C or A_2	$\Delta \bar{V}$ (cm³/mole of product)	K_b/K_m [c]
50,000	1.2648	—	1.2884	−1448	1.5×10^{10}
70,000	1.33	1.32	1.32332	0	1
70,000	1.33	1.32	1.32342	−12	1.21
70,000	1.33	1.32	1.33	−797	4.0×10^5
70,000	1.33	1.32	1.32	+399	1.6×10^{-3}
70,000	1.31	1.32	1.33	−1601	1.8×10^{11}
70,000	1.32	—	1.31	+809.7	2.0×10^{-6}
180,000	1.2648	—	1.2884	−5212	4.2×10^{36}

[a] Adapted from Kegeles et al. (13).
[b] In bimolecular-complex formation, the molecular weight of B was twice that of A. $x_m = 5.7$ cm, $x_b = 7.2$ cm, rotor speed = 60,000 rpm, solution density = 1.05, $\Delta P = 395.8$ atm.
[c] Ratio of association constant at bottom of cell to that at the meniscus.

Factors Governing the Precise Shape of Reaction Boundaries 161

The magnitude of the effect that pressure can have on the velocity-sedimentation behavior of interacting systems has been appreciated only recently. This new insight stems from the studies of Josephs and Harrington (*14*, *15*) on the reversible association of myosin. It is well known that, at low ionic strengths, myosin molecules associate to form long, filamentous macrostructures having features in common with those of native thick filaments of muscle, but usually showing a broad size distribution, dependent, among other things, upon ionic strength. For certain stringent conditions, however, the size distribution may be very sharp. This is the case at ionic strength 0.1–0.2 KCl in the pH range 8–8.5. The molecular parameters of the polymer formed under these conditions are: molecular weight, 50–60×10^6; length, about 6500 Å; diameter, 120 Å; and degree of polymerization, about 83. Detailed sedimentation studies have revealed that the monomer and polymer are in rapid equilibrium interpretable as 83 monomer $\rightleftharpoons$ polymer, with an equilibrium constant[1] at 5°C of about 10^{90} when concentrations are expressed as grams per 100 milliliters. Moreover, the nature of the sedimentation patterns is strikingly dependent upon rotor speed, due to a strong dependence of the equilibrium constant upon hydrostatic pressure, which, in turn, is associated with an unusually large molar-volume change of reaction.

The sedimentation pattern obtained at low rotor speed (9000 rpm) shows two well-resolved peaks—a broad, slowly sedimenting peak and a hypersharp, very rapidly sedimenting one (Fig. 55a). The two peaks constitute a reaction boundary even though the concentration gradient in the region between them is so low as to be virtually undetectable. This is evident from the fact that the area of the slow peak remains constant while that of the fast one increases with increasing concentration to an extent which corresponds to the total increase in concentration. The unusually high resolution of the two peaks is consistent with the predictions of the Gilbert theory for a system in which the single type of polymer molecule is composed of a great number of monomeric units and the equilibrium constant strongly favors association. The theoretical sedimentation pattern computed for the myosin system (Fig. 56) is to be compared with those computed for a tetramerizing protein, Fig. 31 in Chapter III. The distinguishing feature of the theoretical myosin pattern is the extremely small but finite concentration gradient in the region between the two well-resolved peaks constituting the reaction boundary. There is still another important distinction to be made—namely, in this limiting case of

---

[1] Apparently, there is a typographical error by Josephs and Harrington (*15*), in which the value of the equilibrium constant is stated to be 10^{30} when concentrations are expressed in grams per 100 milliliters. The figure 10^{90} was calculated independently either from the data presented by Josephs and Harrington (*14*, Fig. 13) or from the value 10^{482} given therein for the equilibrium constant when concentrations are expressed in moles per liter.

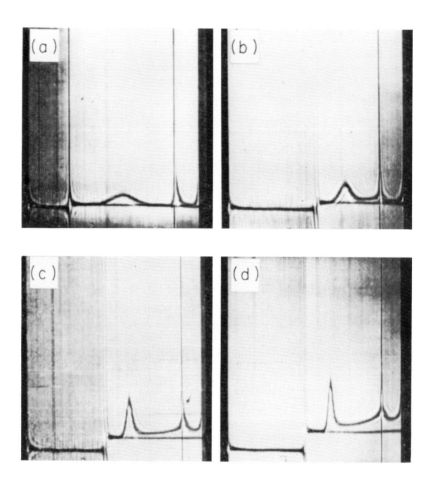

FIG. 55. Effect of rotor speed on the sedimentation pattern of myosin in 0.18 M KCl–2×10^{-3} M barbital, pH 8.3, at 5°C: (a) total protein concentration, 0.4%; time of centrifugation, 18 hr at 9000 rpm; (b) 0.6%, 5 hr at 22,000 rpm; (c) 0.6%, 1.5 hr at 32,000 rpm; and (d) 0.6%, 1 hr at 40,000 rpm. Double-sector cell; sedimentation from left to right. [From Josephs and Harrington (15).]

very large n and K', and in this case only, the area of the slow peak makes a good approximation to the concentration of monomer in the plateau region ahead of the boundary, and that of the fast peak to the concentration of

FIG. 56. Theoretical sedimentation pattern computed from the Gilbert theory for the myosin system, 83 monomer ⇌ polymer. Here, $K' = 10^{90}$ with concentrations expressed as grams per 100 milliliters; total concentration is 0.5 g/100 ml. Sedimentation is from left to right. Compare with the pattern computed for 4 monomer ⇌ polymer in Fig. 30 (Chapter III).

polymer. Moreover, in this limit, the leading edge of the boundary, i.e., the fast peak, sediments with a velocity only slightly less than that of the polymer even at quite low total concentration (Fig. 57). Given this prediction and the

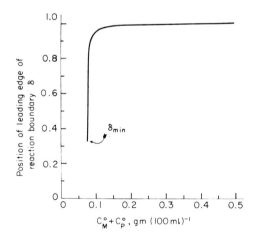

FIG. 57. Theoretical dependence of the position of the leading edge of the bimodal reaction boundary upon total macromolecular concentration for the myosin system, 83 monomer ⇌ polymer; $K' = 10^{90}$ with concentrations expressed as grams per 100 milliliters. Computed from the Gilbert theory. Compare with computations presented in Fig. 31 (Chapter III) for 4 monomer ⇌ polymer.

extreme concentration dependence of the sedimentation coefficient of the asymmetrical polymer molecules, which itself is undoubtedly responsible for the hypersharpness of the fast peak in the experimental patterns, it is not surprising that the observed sedimentation coefficient of the fast peak ($s_{20,w}^\infty = 150$ S) increases continuously and does not pass through a maximum when extrapolated to infinite dilution. The expected maximum in the plot of sedimentation coefficient versus total concentration apparently resides at such a low concentration as not to be realized in practice. Finally, in accord with theoretical expectation, the observed sedimentation coefficient of the slow peak ($s_{20,w}^\infty = 6.5$ S) is characteristic of the myosin monomer.

Now, when identical myosin solutions are sedimented at progressively higher rotor speed, a significant and continuous alteration in the nature of the patterns takes place. This is illustrated in Figs. 55b, c, and d. The salient feature which distinguishes these patterns from the one obtained at lower speed (Fig. 55a) is the marked elevation of the concentration gradient above the baseline both between the two peaks of the reaction boundary and in the region centrifugal to the faster peak. This elevation is seen to increase with rotor speed, suggesting that the monomer–polymer equilibrium is altered as a result of the increasing hydrostatic pressure gradient established throughout the liquid column. That this is in fact the case has been established by two different kinds of experiments.

In the first set of experiments, the hydrostatic pressure on the myosin solution in the centrifuge cell was adjusted by overlaying with increasing thicknesses of mineral oil. In this way, any effect of pressure on the monomer-polymer equilibrium was established before significant sedimentation had occurred at a rotor speed of 40,000 rpm. The resulting sedimentation patterns are displayed in Fig. 58. It will be noted that the elevation of the gradient curve above the baseline between the two peaks and the area of the slower peak increase with increasing pressure. Since, in the limit of very large n and K', the area of the slow peak makes a good approximation to the concentration of monomer in the original equilibrium mixture, it can be concluded that the monomer–polymer equilibrium is a function of hydrostatic pressure and that increasing pressure favors dissociation of the polymer. Supporting evidence was obtained from a second kind of experiment in which the initial concentration profile of the polymer in the centrifuge cell was determined as a function of rotor speed in the absence of applied pressure. In accordance with the thesis of a pressure-dependent equilibrium, concentration gradients of both monomer and polymer should exist throughout the liquid column as soon as the rotor is brought to speed and before any significant mass transport occurs. This is so because the concentrations of monomer and polymer at each level in the column must adjust to satisfy the local value of the equilibrium constant as given by the integrated reaction isotherm, Eq. (260). Thus, the

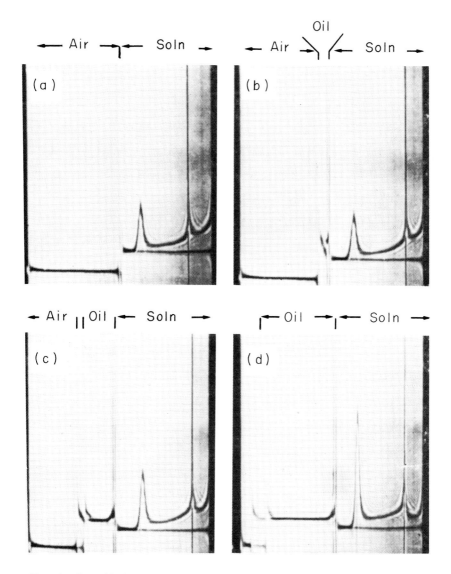

FIG. 58. Effect of hydrostatic pressure on the sedimentation pattern of myosin at constant rotor speed of 40,000 rpm. Varying amounts of mineral oil (density = 0.85 g cm^{-3}), previously equilibrated with dialysate, were added to aliquots of 0.66% myosin solution which had been dialyzed against 0.185 M KCl–2 × 10^{-3} M barbital, pH 8.3. Double-sector cell; sedimentation from left to right. The lower (centrifugal) meniscus at the oil–solution interface is that of the protein solution, and the upper (centripetal) oil–air meniscus corresponds to the sector containing the protein solution. Time of centrifugation, 75 min at 5°C. [From Josephs and Harrington (15).]

monomer concentration should increase continuously from the meniscus to the bottom of the cell, while the polymer concentration should decrease. Since the total concentration remains constant with respect to radial distance, this phenomenon cannot be detected by conventional procedures. Appeal can be made, however, to the fact that the very high molecular weight polymer scatters light far more strongly than does the monomer. Although hardly detectable at the wavelength of light (546 mμ) normally used with the Schlieren optical system, this effect is, by virtue of Rayleigh's scattering law, easily seen at lower wavelength. Figure 59 shows the optical density at 365 mμ as a function of radial distance at four different rotor speeds, obtained from Schlieren photographs taken immediately after reaching speed. It is clear from these results that, at sufficiently high rotor speed, the concentration of polymer does in fact decrease continuously from the meniscus to the bottom of the cell and

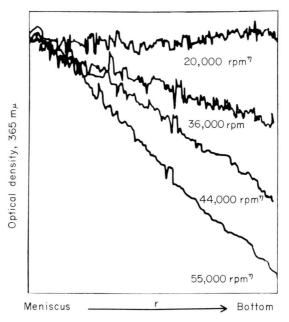

FIG. 59. The effect of rotor speed on the radial optical-density profile of 1.0% myosin in 0.178 M KCl–2 × 10^{-3} M barbital, pH 8.3, at 5°C in a 30-mm single-sector cell. Optical density is in arbitrary units and was obtained from microdensitometer tracings of Schlieren photographs taken with ultraviolet (365 mμ) light immediately upon reaching speed and before any significant mass transport had occurred. At 52,000 rpm, the extinction change across the photographic plate from meniscus to cell bottom is about 0.9 optical density (OD) unit. Identical results were obtained irrespective of whether the rotor speed was raised from 20,000 to 52,000 rpm or first raised to 52,000 rpm and then decreased to 20,000 rpm. [Adapted from Josephs and Harrington (15).]

that its absolute concentration gradient increases markedly with increasing rotor speed.

Finally, the exquisite sensitivity of the reversible polymerization of myosin to pressure indicates an unusually large molar-volume change of reaction. Its magnitude has been evaluated in the usual manner from plots of $\log K'$ vs. P to be about $+380$ cm^3 per monomer unit, or $+32$ liters per mole of polymer. Elucidation of the molecular mechanism of this effect and of the influence of ionic strength and other environmental factors upon the size distribution of myosin polymers formed *in vitro* may well have important implications for the developmental biology of muscle cells. In addition, many scientists believe that such studies may provide the insight required for the control of primary muscular disease. In any case, they obviously have important implications for the ultracentrifugal characterization of interacting systems.

The theory of the effect of pressure on the velocity-sedimentation behavior of reversibly reacting systems has been elaborated by Kegeles and his co-workers (*13*). Computations were made for dimerization and bimolecular complex formation under the assumption of instantaneous reestablishment of chemical equilibrium during differential transport of the reacting species. The effect of pressure was introduced into the numerical solution of the exact conservation equations by calculating the equilibrium constant at each position using Eq. (260), the concentration of each species being required to satisfy the local value of the equilibrium constant. The results of these calculations are of considerable fundamental importance, since they reveal that, in the case of pressure-sensitive equilibria, the sedimentation process can generate inverted gradients of total macromolecular concentration either within the reaction boundary or at its leading edge. In other words, the total gradient can change sign and become depressed below the base line in these regions, which means that the density of the solution decreases with increasing radial distance. Clearly, such a situation would be an unstable one in ordinary sedimentation-velocity experiments and would result in continuing convective disturbance. In that event, quantitative, and possibly even qualitative, interpretation of the sedimentation patterns becomes difficult or impossible in the absence of a stabilizing density gradient.

Theoretical sedimentation patterns calculated for two monomer–dimer systems which differ in their molecular and chemical parameters are displayed in Fig. 60. Consider first the case in which pressure favors formation of dimer (Fig. 60a). We see that, while the gradient of monomer concentration shows both a positive and a negative peak, the gradient of total concentration is everywhere positive in this particular example. For other values of the association constant, however, it is possible for the patterns to show an inverted gradient at the leading edge of the reaction boundary. This is illustrated in

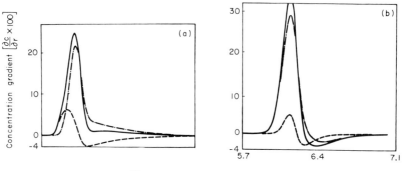

FIG. 60. The effect of pressure on the velocity sedimentation of dimerizing systems: (———) total; (– – –) monomer; (– · –) dimer. Theoretical sedimentation patterns computed for two different dimerizing proteins: (a) pressure favors dimer, monomer density is 1.32, dimer density is 1.33, $\Delta \bar{V} = -798$ cm^3 (mole of dimer)$^{-1}$, association constant is 1; (b) pressure favors monomer, monomer density is 1.32, dimer density is 1.31, $\Delta \bar{V} = +810$ cm^3 (mole of dimer)$^{-1}$, association constant is 10. Monomer molecular weight is 70,000; sedimentation coefficients are 4 S and 6 S; solution density is 1.05; 60,000 rpm. [From Kegeles et al. (13).]

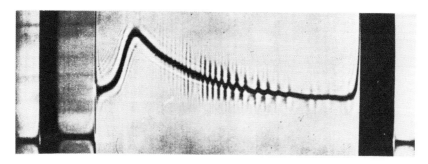

FIG. 61. An ultracentrifuge pattern illustrating convection during a sedimentation-velocity experiment. The sedimenting material is serum albumin in which most of the disulfide bonds have been reduced and the resulting sulfhydryl groups reacted with iodoacetic acid. The reacted protein aggregates considerably. [From Schachman (16).]

Fig. 60b for the case where pressure favors dissociation of the dimer. Under these circumstances, convection will occur in practice unless a stabilizing density gradient is present. The convective disturbance may be manifested in the Schlieren pattern as small pips centrifugal to the front, erosion of the boundary, or, in certain instances, hypersharpening of the boundary. Apparently, for one reason or another, convection was not a complicating factor in

Factors Governing the Precise Shape of Reaction Boundaries 169

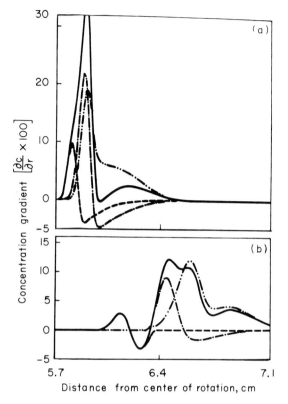

FIG. 62. Theoretical velocity sedimentation patterns for a reacting system of the type A+B ⇌ C with pressure favoring complex formation: (a) Early time of sedimentation; (b) later time. (———) total; (- - -) A; (- · -) B; (- · · -) C. Density of reactants and complex are 1.31, 1.32, and 1.33; $\Delta \bar{V} = -1601$ cm³ (mole of complex)$^{-1}$; and solution density, 1.05. Molecular weights are 70,000, 140,000, and 210,000; sedimentation coefficients are 4, 6 and 8 S; association constant is 1; 60,000 rpm. [From Kegeles et al. (13).]

the experiments of Josephs and Harrington (14, 15) on the association of myosin, although it should be pointed out that convective disturbances are often difficult to detect (16). On the other hand, convection has been observed with other associating–dissociating proteins (16, 17). An extreme example is shown in Fig. 61.

Sedimentation patterns have also been calculated for reacting systems of the type A + B ⇌ C for the case where pressure favors complex formation. The time course of development of the reaction boundary is illustrated in Fig. 62. During early stages of sedimentation, the total gradient is everywhere positive, although those of the reactants can become negative (Fig. 62a). As sedimentation proceeds, however, the total gradient becomes inverted between the two slowest-sedimenting peaks in the pattern, which, rigorously, is a single multimodal reaction boundary. Here, too, convective disturbances can be expected to occur between the two slowest peaks in the absence of a stabilizing density gradient. The prediction that a system of the type $A + B \rightleftharpoons C$ can give rise to a sedimentation pattern showing five peaks (four positive and one negative) is indeed startling. When the association constant or the volume change of reaction was decreased in value, the number of positive peaks decreased to three or two, and the negative peak diminished in area, but rarely disappeared.

Thus, we see that the results of both experimental and theoretical investigations attest to the fact that the pressure generated in the ultracentrifuge cell at high rotor speed can have a profound effect on the velocity-sedimentation behavior of reacting systems. Moreover, the change in volume which can give rise to significant pressure dependence lies well within the present accuracy of routine partial specific volume determinations. For example, the absolute difference between the partial specific volumes of monomeric myosin and its polymer is only $6 \pm 1.2 \times 10^{-4}$ cm^3 g^{-1}, but, because of the high molecular weight of the monomer and the high degree of polymerization, the molar-volume change of reaction (monomer → polymer) is +32 liters per mole of polymer. Even in the case of the theoretical dimerization reaction illustrated in Fig. 60b, the value of +810 cm^3 $mole^{-1}$ assigned to the molar-volume change of reaction corresponds to an absolute difference of only 6×10^{-3} cm^3 g^{-1} between the partial specific volumes of monomer and dimer. Clearly, high-speed sedimentation-velocity experiments must be interpreted with great caution, since, in general, one simply cannot tell *a priori* whether or not a given system will be pressure sensitive. Several measures have been suggested (*13*) in order to avoid misinterpretation. These include (1) introduction, at least in exploratory experiments, of stabilizing density gradients of an inert solute, lest unobserved convective disturbances negate the possibility of meaningful interpretation; (2) variation of rotor speed; and (3) application of hydrostatic pressure by overlaying the aqueous solution with oil, as was done in the studies on myosin polymerization. The latter is essential in order to distinguish between pressure-dependent reactions and other types of interaction which may also show rotor-speed dependence; e.g., kinetically controlled processes and, as will become clear shortly, macromolecule–small-molecule interactions. In addition, there are the previously stressed precautionary procedures of varying the macromolecular concentration over a wide range, fractionation using partition cells, and acquisition of independent physicochemical information such as might be obtained from light-scattering, electrophoretic, and gel filtration measurements. Finally, molecular weights cannot be computed with confidence from sedimentation coefficients which have not been demonstrated to be independent of rotor speed. As illustrated in Table

III, association constants which are negligibly small at 1 atm pressure may give rise, nevertheless, to strong interactions in the ultracentrifuge cell at high speed. Accordingly, molecular weights are more suitably determined by low-speed equilibrium sedimentation or by the low-speed Archibald method.

ZONE TRANSPORT OF REACTING SYSTEMS

Up to this point, our discussion has been concerned chiefly with moving-boundary transport of reacting systems. Mention has been made, however, of the fact that the conclusions reached in the case of macromolecular isomerization are also valid for zone separation processes such as partition chromatography and zone electrophoresis. Thus, for example, an isomerizing protein can show as many as three spots on paper chromatography (9). Theoretical investigations have also been made into the zone transport of systems undergoing other types of interaction—namely, association–dissociation, bimolecular complex formation, and macromolecule–small-molecule interactions. Discussion of the latter properly awaits the full account of this type of interaction given in the next section. It suffices here to state that a single macromolecule, which interacts reversibly with a small uncharged constituent of the solvent with concomitant change in mobility, can give two electrophoretic zones despite instantaneous establishment of equilibrium. Rigorously speaking, the two zones actually constitute a single bimodal reaction zone analogous to a bimodal reaction boundary. Thus, while the concentration of macromolecule in the region separating the two zones may become very small during their differential migration along the supporting medium, it always remains finite. In practice, however, it is virtually impossible to distinguish such zone patterns from those shown by a mixture of two non-interacting macromolecules, and conclusions concerning heterogeneity must rest upon fractionation experiments. On the other hand, these calculations are providing the theoretical basis for the application of zone-separation processes to the study of macromolecule–small-molecule interactions.

The exact theory of zone transport of associating–dissociating and reversibly complexing systems in which equilibration is instantaneous has been elaborated by Bethune and Kegeles (18, 19). Since their calculations were made for the case of countercurrent distribution (20), the results are displayed in Figs. 63–66 as plots of the total weight of macromolecule contained in each successive tube of equal volume in the countercurrent distribution train versus the reduced position coordinate, r/n, where r is the tube number and n is the number of transfers. The analytical functions which describe the shape of the zones were solved numerically. When the number of extractions is small, these functions are not continuous, but after a large number of transfers, the discontinuous functions may be replaced by continuous ones. Accordingly, the analogy is readily drawn among countercurrent distribution, partition chromatography, and zone electrophoresis. Thus, for zone electrophoresis, the weight of material in each tube transforms into the total amount of macromolecule at position $r \to x$ along the supporting medium after the time of electrophoresis $n \to t$.

In the case of reversible association, $nM \rightleftharpoons P$, the theoretical zone patterns (*18*) show only a single peak for both dimerization and trimerization (Fig. 63).

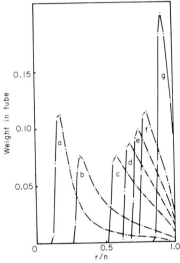

FIG. 63. Theoretical countercurrent distribution patterns for a trimerizing system, illustrating the effect of varying the association constant, which decreases from (a) to (g). [Adapted from Bethune and Kegeles (*18*).]

The peak is quite asymmetrical in both instances if significant quantities of monomer and polymer coexist, and is hypersharp along that edge at which the polymer would occur if it were to exist alone. In the zone electrophoretic analogy, this means that the leading edge of the zone will be hypersharp and the trailing edge diffuse when the mobility of the polymer is greater than that of the monomer, and vice versa when the monomer has the greater mobility. This, in turn, is analogous to the situation encountered in the preceding chapter where comparison was made between ascending and descending fronts in moving-boundary electrophoresis of reversibly polymerizing systems. It is esthetically pleasing that accurate numerical solution of exact equations substantiates the analytical and physical reasoning used in that discussion. There is still another important feature of some of the zone patterns of trimerizing, but not dimerizing, systems—namely, an inflection point in the diffuse edge of the zone which gives the impression of incipient resolution which, in fact, never occurs. This feature is clearly discernible in Figs. 63c–f, and in practice might lead to an incorrect conclusion of inherent heterogeneity if fractions taken from selected positions in the zone were not reprocessed. Since the position at which the zone occurs is a function of initial concentration, a change in position upon reprocessing without an intervening concentrating step would be indicative of a reaction. If the concentration of the fraction taken from any one position were reconstituted to the original concentration prior to reprocessing, it should show the same pattern as the unfractionated material. Simple dilution or concentration of the original material would have similar effects.

Turning our attention to bimolecular complex formation, $A + B \rightleftharpoons C$, we discern six cases, depending upon the relative mobilities of A, B, and C. Bethune and Kegeles (19) have made computations for several of these, but we shall restrict ourselves to the case most frequently encountered in practice, that in which the mobility of the complex is intermediate between those of the reactants, $v_A > v_C > v_B$. Such a system can show one, two, or three peaks, depending upon the value of the equilibrium constant and the number of transfers in countercurrent distribution, the length of the chromatographic column, or the time of zone electrophoresis. This behavior is illustrated in Figs. 64–66. If the equilibrium constant is very large, as in Fig. 64, a single peak corresponding to the complex is observed during early stages of trans-

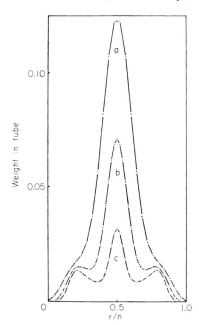

FIG. 64. Theoretical countercurrent distribution pattern for bimolecular complex formation, $A + B \rightleftharpoons C$, with a large equilibrium constant: (a) 25 transfers; (b) 50 transfers; (c) 100 transfers. [Adapted from Bethune and Kegeles (19).]

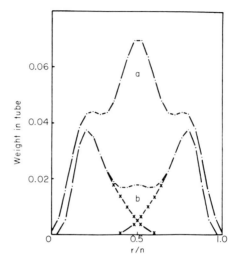

FIG. 65. Theoretical countercurrent distribution pattern for bimolecular complex formation $A + B \rightleftharpoons C$, with an equilibrium constant of intermediate value: (a) 25 transfers; (b) 50 transfers. ($\times$—$\times$) Amount of uncombined A or B in successive tubes of the countercurrent distribution train. [Adapted from Bethune and Kegeles (19).]

FIG. 66. Theoretical countercurrent distribution pattern for bimolecular complex formation, $A + B \rightleftharpoons C$, with a small equilibrium constant: (a) 25 transfers; (b) 50 transfers. (The points denoted by $\times$ are irrelevant to the discussion.) [Adapted from Bethune and Kegeles (19).]

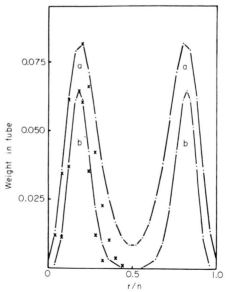

port; but, as the migrating zone spreads by diffusion, the complex dissociates in the dilute leading and trailing edges. Once the newly formed B is left behind, in the absence of A or C, and A moves out ahead, without B or C, they are transported independently. Consequently, the zone develops three peaks, the peripheral ones corresponding to A and B. One would expect that, eventually, all of C would dissociate due to dilution, in which case, A and B would become completely separated. However, as long as three peaks coexist, they cannot resolve completely, since the intermediate portion of the pattern may be termed a reaction zone within which the equilibrium $A + B \rightleftharpoons C$ continually readjusts during differential transport of the three species. The growth of the A and B peaks during transport is quite obvious in Fig. 65 for intermediate values of the equilibrium constant. Finally, for small value of the equilibrium constant (Fig. 66) two well-resolved zones corresponding to pure A and B evolve quite rapidly.

A striking example of considerable importance for biological control mechanisms is provided by tryptophan synthetase from *E. coli*. In this system, two separable proteins A and B must be combined to give an enzyme which catalyzes the transformation of indole and serine to tryptophan. Chromatography of the original extract of tryptophan synthetase on DEAE cellulose (*21*) gives two major peaks of activity (and an additional small one) when protein B is added to the assay system, two when protein A is added, and one when there is no addition of A and B. The latter peak, which contains the complex, is intermediate in position between the A and B peaks of the chromatogram. When rechromatographed, it reproduces the pattern of the original extract (except for the small peak of A activity noted above). Moreover, when pure A and B are isolated, each chromatographs as a single peak, but, when admixed, they give a pattern resembling that of the unfractionated system. The maximum time for recombination of A and B is less than 2 min, but the rate of dissociation has not been determined.

Certain of the theoretical conclusions reached above may be carried over to band sedimentation and zone velocity sedimentation in a preformed density gradient, but only when the equilibrium constant is insensitive to pressure. The theory has been extended by Kegeles *et al.* (*13*) to include the band sedimentation of pressure-sensitive systems. A most interesting result was obtained for reversible dimerization in which dimer formation is favored by pressure. The band skews backwards, but does not resolve as it moves down the centrifuge cell with an ever-increasing sedimentation coefficient due to progressive accentuation of the dimer in the pressure gradient. By the same token, such a system would show an increase in sedimentation coefficient with increase in rotor speed. For bimolecular complex formation ($v_A < v_B < v_C$) with pressure favoring the complex, the initial band resolves into a slow band corresponding to pure A and a faster, barely bimodal band which contains both B and C

(Fig. 67). Offhand, this is a rather unexpected result; however, upon reflection, we see that it follows directly from the fact that the complex is stabilized by pressure as it moves down the cell and accordingly cannot dissociate completely. The correct sedimentation coefficient of A can be determined from the pattern. Those of B and C cannot, even if specific methods are used to locate the species, since the zone of each no longer moves at the assigned rate. This finding has far-reaching implications for the many applications of zone velocity sedimentation in a density gradient to problems in molecular biology. In this connection, it should be mentioned that the hydrostatic pressure developed in the centrifuge cell at high rotor speeds can also have important effects on the density-gradient equilibrium sedimentation of interacting macromolecules (*13, 22*).

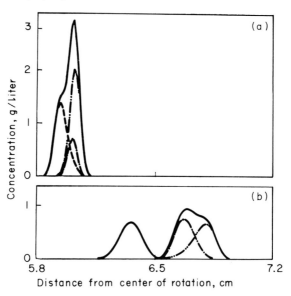

FIG. 67. Theoretical band-sedimentation pattern for bimolecular complex formation $A + B \rightleftharpoons C$, illustrating the effect of the pressure gradient generated in the ultracentrifuge cell when pressure favors complex formation: (a) early time of sedimentation; (b) later time. (——) Total; (– – –) A; (– · –) B; (– · · –) C. Molecular weights of A, B, and C are 70,000, 140,000, and 210,000, respectively; sedimentation coefficients are 4, 6, and 8 S; densities are 1.31, 1.31, and 1.32; $\Delta \bar{V} = -1218$ cm^3 mole^{-1}; solution density is 1.0; association constant is 0.1; 60,000 rpm. [From Kegeles *et al.* (*13*).]

INTERACTION OF MACROMOLECULES WITH SMALL MOLECULES

Once more, we return to the important subject of the interaction of macromolecules with small molecules. Many of the current ideas concerning the molecular mechanisms of cellular regulatory processes derive from a large body of knowledge concerning the binding of small molecules and ions by a variety of proteins and the effect of such interactions upon macromolecular conformation and state of aggregation. An example is the Monod–Wyman–Changeux model of allostery (*23*). It has also been proposed that, in some cases, multiple molecular forms of a given enzyme (isozymes) may differ from each other in conformation rather than in primary structure (*24*). Such conformational variants could conceivably be generated by interactions with small molecules. It may be relevant that auxotropic mutants of glutamic acid dehydrogenase of *Neurospora crassa* are activated *in vitro* by exposure to glutamate, succinate, malate, or acetate (*25, 26*). Enzyme induction is another phenomenon which will presumably find its explanation in such terms (*27–29*). Thus, isopropyl thiogalactaside and other small molecules which induce the synthesis of β-galactosidase in *Escherichia coli* are thought to exert their action by reacting with and thereby inactivating the repressor molecule, which, in turn, is evidently a protein.

As we saw in Chapter II, electrophoresis is a useful method for characterizing the interaction of proteins with small ions. In general, the number and intrinsic affinity of combining sites on the protein molecule can be determined by applying the concepts of the weak-electrolyte moving-boundary theory in conjunction with the law of mass action. Briefly, the concentration of unbound small ion in a protein–small-ion mixture of known overall composition is calculated from a knowledge of the constituent mobilities of the protein and small ion. To determine the constituent mobility of the protein, for example, it is merely necessary to make a moving-boundary experiment in which the initial boundary is formed between a buffered solution of protein and the small ion and buffer containing the small ion. Both the ascending and descending boundaries are reaction boundaries within which the protein–small-ion equilibria are continually readjusting during differential transport of the various complexes. However, for reasons cited previously, only the descending boundary gives the constituent mobility of the protein in the initial equilibrium mixture. This approach owes its success, in part, to the fact that the descending boundary (as well as the ascending one) is generally unimodal, so that there can be no interpretative ambiguity. This is so because strong concentration gradients of the small ion produced by reequilibration cannot be maintained within the boundary against rapid electromigration of the small ion. If strong gradients were maintained, the reaction boundary might well be multimodal, as in the case of interaction of proteins with electrically neutral molecules of undissociated buffer acid. Quantitative characterization of such systems requires a more penetrating analysis than provided by the weak-electrolyte moving-boundary theory, although the latter may serve as an aid in establishing the nature of the interaction. [See Cann and Goad (*30*, footnote 7) for discussion of the limitations on the application of the weak-electrolyte moving-boundary theory to protein–neutral-molecule interactions.] The remainder of this chapter is devoted to the theory of electrophoresis of systems in which the macromolecule interacts with a small, neutral molecule, with concomitant change in mobility of the macromolecule due to a change in its net charge without significant change in frictional coefficient, and extension of the theory to include ultracentrifugation of allosteric types of association–dissociation interactions.

Longsworth and Jacobsen (*31*) were the first to note that the moving-boundary electrophoretic patterns of some proteins at pH values acid to their isoelectric points are suggestive of interactions. Cann and his co-workers (*30, 32–42*) have pursued this matter and observed that a variety of proteins display nonenantiographic electrophoretic patterns in acidic media containing acetate, formate, or other carboxylic acid buffers. The nature of the patterns is critically dependent in a reversible fashion upon the concentration of buffer in the solvent medium, but is insensitive to ionic strength. The behavior typically shown by proteins in media containing varying concentration of

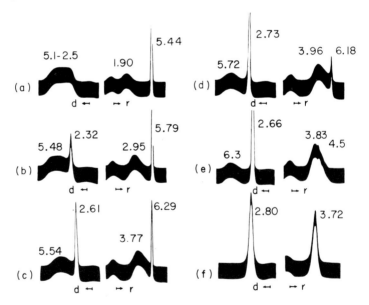

FIG. 68. Moving-boundary electrophoretic patterns of ovalbumin in media containing varying concentration of formate buffer (Naf–Hf), pH 4.0: buffer composition (molar concentration) (a) 0.040 Naf–0.023 Hf; (b) 0.020 Naf–0.012 HF–0.020 NaCl; (c) 0.014 NaF–0.0081 HF–0.026 NaCl; (d) 0.012 Naf–0.0069 Hf–0.028 NaCl; (e) 0.010 Naf–0.0058 Hf–0.038 NaCl; (f) 0.002 NaF–0.0012 HF–0.038 NaCl. Essentially the same behavior is shown in media containing acetate (NaAc–HAc) or other carboxylic acid buffers. Thus, for example, the patterns in 0.01 M NaAc–0.05 M HAc–0.03 M NaCl (*33*, Fig. 1) are virtually the same as those in 0.040 M Naf–0.023 M Hf. [From Cann and Phelps (*33*).]

buffer is illustrated in Figs. 68 and 69. The nonenantiography is quite bizarre and not at all like that shown by mixtures of noninteracting proteins. Moreover, increasing the buffer concentration at constant pH and ionic strength results in progressive and characteristic changes in the patterns, notably in the appearance and growth of fast-moving peaks at the expense of slow ones. Fractionation and other measurements have established that the peaks in a given pattern constitute a reaction boundary, modified in some nistances by mild convective disturbances, and are in no sense indicative of inherent heterogeneity. Evidence has been advanced to support interpretation of these observations in terms of reversible complexing of the protein molecules with undissociated buffer acid, with concomitant subtle structural changes which increase the net positive charge on the protein, but do not change its frictional coefficient significantly.

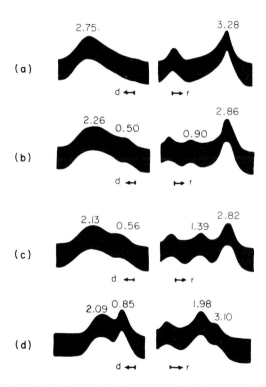

FIG. 69. Moving-boundary electrophoretic patterns of bovine γ-pseudoglobulin in media containing varying concentration of acetate buffer, pH 5.7: buffer composition (molar concentration) (a) 0.04 NaAc–0.0040 HAc; (b) 0.01 NaAc–0.010 HAc–0.03 NaCl: (c) 0.005 NaAc–0.00050 HAc–0.035 NaCl; (d) 0.002 NaAc–0.00020 HAc–0.038 NaCl. [From Cann and Phelps (33).]

A theoretical basis for this interpretation is provided by the theory of electrophoresis of interacting systems described below. These computations account for the essential features of the electrophoretic behavior of proteins in acidic media containing varying concentrations of carboxylic acid buffer. Furthermore, the theory predicts that a single macromolecule reacting reversibly with a small, uncharged constituent of the solvent medium can give enantiographic moving-boundary patterns showing two well-resolved peaks or a zone pattern showing two zones. This is so despite instantaneous establishment of equilibrium. The implications for conventional electrophoretic analysis are evident. Further, as with all theoretical advances, the new understanding provided by these calculations suggests new and interesting applications of electrophoresis and other transport methods to biological problems such as the molecular mechanisms of cellular control processes.

Formulation of Theory. A theory of electrophoresis has been developed (30, 40) for interacting systems of the type

$$P + nHA \rightleftharpoons P(HA)_n \tag{261}$$

where P represents a protein molecule or other macromolecular ion in solution and $P(HA)_n$ its complex with n moles of a small uncharged constituent HA of the solvent medium, for example, undissociated buffer acid. It is assumed that the complex migrates in an electric field with an electrophoretic mobility μ_2 that differs from the mobility μ_1 of the uncomplexed protein. The molar concentrations of P, $P(HA)_n$, and HA are designated as C_1, C_2, and C_3, and their diffusion coefficients as D_1, D_2, and D_3, respectively. The transport-interconversion equations for the three species are

$$\frac{\partial C_1}{\partial t} = D_1 \frac{\partial^2 C_1}{\partial x^2} - \mu_1 E \frac{\partial C_1}{\partial x} + R_1 \tag{262a}$$

$$\frac{\partial C_2}{\partial t} = D_2 \frac{\partial^2 C_2}{\partial x^2} - \mu_2 E \frac{\partial C_2}{\partial x} - R_1 \tag{262b}$$

$$\frac{\partial C_3}{\partial t} = D_3 \frac{\partial^2 C_3}{\partial x^2} + nR_1 \tag{262c}$$

in which R_1 is the net rate of formation of P due to the reversible reaction (261). These three equations constitute the set of differential equations whose solution gives the desired electrophoretic patterns. [Equation (262c) does not include a force term, since the uncharged small molecule does not move under the influence of the electric field.]

Interaction of Macromolecules with Small Molecules 181

The first mathematical manipulation of Eqs. (262a–c) is to recast them as follows: Addition of Eqs. (262a) and (262b) yields

$$\frac{\partial(C_1 + C_2)}{\partial t} = D_1 \frac{\partial^2 C_1}{\partial x^2} - \mu_1 E \frac{\partial C_1}{\partial x} + D_2 \frac{\partial^2 C_2}{\partial x^2} - \mu_2 E \frac{\partial C_2}{\partial x} \qquad (263a)$$

which is a statement of conservation of total protein under the action of electrophoretic transport and diffusion. The conservation equation for total small molecule is obtained by multiplying both sides of Eq. (262b) by n and adding the product to Eq. (262c):

$$\frac{\partial(nC_2 + C_3)}{\partial t} = nD_2 \frac{\partial^2 C_2}{\partial x^2} - n\mu_2 E \frac{\partial C_2}{\partial x} + D_3 \frac{\partial^2 C_3}{\partial x^2} \qquad (263b)$$

Now a third equation must express the effect of reaction (261) on the concentrations. The calculations have been made for systems in which the reaction rate is very much larger than the rates of diffusion and electrophoresis, and local equilibrium at every instant has been assumed. Accordingly,

$$C_2 = KC_1 C_3^n \qquad (263c)$$

where K is the equilibrium constant for reaction (261). The new set of equations (263a–c), which makes an explicit statement as to the rates of reaction, must now be solved.

If the concentration of HA is very large compared with that of the protein, the reaction cannot appreciably affect C_3, and it will be constant to a good approximation. Then Eq. (263c) tells us that C_2 is just equal to C_1 multiplied by the constant factor KC_3^n, in which case, Eq. (263a) can be written in a form in which only the variable $C_1 + C_2$ appears with concentration-averaged diffusion coefficient and mobility, namely,

$$\frac{\partial(C_1 + C_2)}{\partial t} = \bar{D} \frac{\partial^2(C_1 + C_2)}{\partial x^2} - \bar{\mu} E \frac{\partial(C_1 + C_2)}{\partial x}$$

$$\bar{D} = \alpha_1 D_1 + \alpha_2 D_2 \quad \text{and} \quad \bar{\mu} = \alpha_1 \mu_1 + \alpha_2 \mu_2 \qquad (264)$$

$$\alpha_1 = \frac{1}{1 + KC_3^n} \quad \text{and} \quad \alpha_2 = \frac{KC_3^n}{1 + KC_3^n}$$

The form of this equation is the same as the forced-diffusion equation [Eq. (19) in Chapter I] for a single, noninteracting macromolecular ion. Accordingly, in this limit, the two species behave as a single macromolecular ion with average transport properties, and the electrophoretic patterns show a single Gaussian-shaped boundary or a single zone. It is also of some interest to note the analogy to rapid isomerization, Eq. (85) in Chapter II.

In the general case, however, we must deal with a set of nonlinear partial differential equations, and it is necessary to resort to numerical methods. The reader is referred to the original publication (30) and to Chapter V for mathematical details. Suffice it to say that after transformation of the conservation equations to a moving frame of reference, the corresponding finite-difference equations were solved on a high-speed electronic computer using initial and boundary conditions corresponding to either moving-boundary or zone electrophoresis. One can regard this procedure as a matter of following the motion of molecules during a short time interval as they pass into and out of a number of consecutive, thin cross-sectional slices of the electrophoresis column. At the same time, the equilibrium condition is imposed within each slice. This is done repeatedly to obtain the development of the electrophoretic pattern with time.

The concentrations of P and P(HA)$_n$ in the initial equilibrium solution are designated as C_{10} and C_{20}, respectively. The concentration of HA is generally expressed as pHA = $-\log C_3$, although the symbol C_{30} is used to indicate initial concentration. Except where indicated, the following values for the several molecular parameters were used in the computations presented here: $\mu_1 = 3 \times 10^{-5}$ cm^2 sec^{-1} V^{-1}, $\mu_2 = 5 \times 10^{-5}$, $D_1 = D_2 = 3.14 \times 10^{-7}$ cm^2 sec^{-1}, and $D_3 = 5.5 \times 10^{-6}$. The latter is typical of small molecules having molecular weight of the order of 100. The quantity E was assigned a value of 10 V cm^{-1}; and, unless otherwise indicated, $C_{10} + C_{20} = 1.4 \times 10^{-4}$ M, which would correspond to a 1% solution of a protein with molecular weight of approximately 70,000. The results of the computations are displayed graphically as (1) plots of $\delta(C_1 + C_2)/\delta x$ against position in the electrophoresis column for theoretical moving-boundary electrophoresis patterns (in Figs. 70, 74, and 75, plots of C_1 and C_2 against position are also presented), and (2) plots of $(C_1 + C_2)$ versus position for theoretical zone-electrophoretic patterns. Both are accompanied by plots of pHA against position. Migration velocities per unit field, $10^5 \times v$ cm^2 sec^{-1} V^{-1}, are shown above or beside corresponding peaks or zones.

Results for Moving-Boundary Electrophoresis. Representative results of the computations for moving-boundary electrophoresis (30) are presented in Figs. 70–75. The several different sets of conditions chosen for these computations approximate those for which protein–buffer acid interactions have been encountered experimentally. (Note that the value of C_{30} for which $C_{20}/C_{10} = 1$ is equal to the value of $K^{-1/n}$.) Those in Figs. 70 and 71 correspond to γ-pseudoglobulin at pH 5.7, and Figs. 72 and 73 correspond to ovalbumin and a variety of other highly purified proteins at pH 4.

Let us focus attention for the moment upon Fig. 70, which shows the time course of development of the theoretical electrophoretic patterns for $C_{30} =$

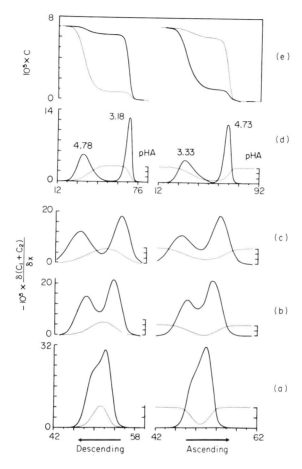

FIG. 70. Time course of development of theoretical moving-boundary electrophoretic patterns for $K^{1/n} = 10^4$, $n = 3$, and $C_{20}/C_{10} = 1$. Four lower graphs: (———) gradient curve; (- - -) pHA. Topmost curve: (———) C_1; (- - -) C_2. Time of electrophoresis: (a) 1.69×10^3 sec; (b) 3.37×10^3 sec; (c) 5.06×10^3 sec; (d) and (e) 2.19×10^4 sec. The abscissas in this and all other figures displaying moving-boundary patterns are $10x + 50 \pm 4 \times 10^{-3}t$ for descending and ascending patterns, respectively. pHA scale for (d): descending, 4.0–4.3; ascending, 3.7–4.0. [From Cann and Goad (30).]

10^{-4} M, $C_{20}/C_{10} = 1$, and $n = 3$. It is strikingly apparent that the theory predicts resolution of the patterns into two moving peaks. This, despite instantaneous reestablishment of equilibrium during differential migration of P and P(HA)$_3$. It is also evident that resolution occurs because of changes in the concentration of HA accompanying reequilibration and maintenance of the resulting concentration gradients of the electrically neutral molecule

The macromolecular concentration curves reveal that the peaks in the patterns correspond to different equilibrium compositions and not simply to P or $P(HA)_3$. Nor do the peaks migrate with the mobilities of P and $P(HA)_3$. Furthermore, the calculations show that a homogeneous phase cannot be generated between the two peaks. Although the protein gradient eventually becomes very small between the peaks, it never vanishes. In other words, the patterns display bimodal reaction boundaries. An important feature of the ascending and descending patterns is their marked nonenantiography, arising solely from the interaction. While, in the ascending pattern, the fast-moving peak is sharp and the slow-moving one is broad, in the descending pattern, the slow-moving peak is the sharp one and the fast peak is broad. Such nonenantiography is characteristic of experimental patterns of proteins in acidic media containing appropriate concentration of carboxylic acid buffer (e.g., Fig. 68c).

The areas enclosed by the peaks in the particular patterns shown in Fig. 70 are approximately proportional to C_{10} and C_{20}, but the agreement is fortuitous. In general, there is no simple relationship between the areas of the peaks and the concentrations of P and $P(HA)_3$ in the initial equilibrium mixture. This is illustrated in Fig. 71 by a set of patterns computed for in-

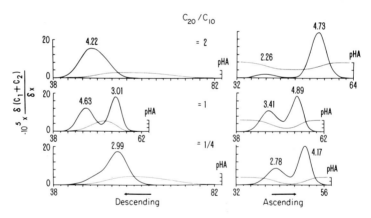

FIG. 71. Theoretical moving-boundary electrophoretic patterns for $K^{1/n} = 10^4$, $n = 3$, and values of C_{20}/C_{10} shown. (——) Gradient curve; (– – –) pHA. Virtually the same results were obtained for $n = 1$, the only differences being quantitative ones. Time of electrophoresis: $C_{20}/C_{10} = \frac{1}{4}$ and 2, 5.74×10^3 sec; $C_{20}/C_{10} = 1$, 5.06×10^3. [From Cann and Goad (30).]

creasing value of C_{20}/C_{10} (i.e., increasing C_{30}) and is understandable in terms of the fact that the peaks correspond to different equilibrium compositions rather than single components.

Comparison of Fig. 71 with Fig. 69 shows that the theoretically computed patterns reproduce quite closely the behavior of bovine γ-pseudoglobulin in media containing varying concentration of acetate buffer at pH 5.7. Since results similar to those illustrated for $n = 3$ have also been obtained for $n = 1$, it is reasonable to ascribe the electrophoretic behavior of γ-pseudoglobulin to the binding of 1–3 moles of acetic acid, with concomitant increase in mobility.

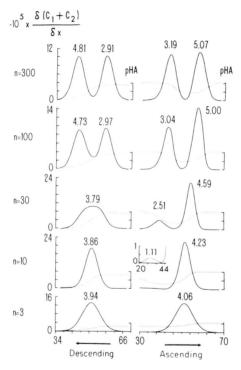

FIG. 72. Theoretical moving-boundary electrophoretic patterns for $K^{1/n} = 10^2$, $C_{20}/C_{10} = 1$, and values of n shown opposite patterns. (———) Gradient curve; (– – –) pHA. Time of electrophoresis: $n = 3$, 1.72×10^4 sec; $n = 10$, 6.92×10^3; $n = 30, 100,$ and 300, 7.17×10^3. pHA scale for $n = 30$: descending, 2.00–2.02; ascending, 1.98–2.00. [From Cann and Goad (30).]

The computations also predict all of the essential features of the electrophoretic behavior in carboxylic acid-containing media at pH 4. Compare, for example, the theoretical patterns presented in Fig. 72 for $n = 30$ with the experimental ones shown in Fig. 68a. Indeed, the particular type of nonenantiography displayed by these patterns (two well-resolved ascending peaks but only a single, broad descending one) is shown typically by a variety of proteins in media containing 0.01 M sodium acetate buffer. In addition, the theory accounts for (1) the progressive changes induced in the patterns by systematic variation of buffer concentration [see theoretical patterns presented in Figs. 7 and 8 or Cann and Goad (30)]; (2) the observed changes in pH across the several peaks in the patterns; (3) the dependence of the electrophoretic patterns upon protein concentration; and (4) the mild convective disturbances often observed in ascending electrophoretic patterns. The latter two points require further comment.

Theoretical ascending patterns computed for decreasing values of $C_{10} + C_{20}$, other things being held constant, are presented in Fig. 73. These results

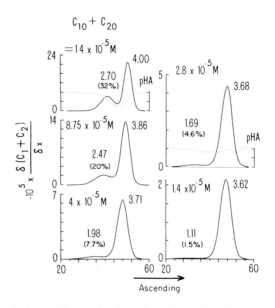

FIG. 73. Theoretical ascending moving-boundary electrophoretic patterns for $K^{1/n} = 10^2$, $n = 30$, $C_{20}/C_{10} = \frac{1}{4}$, and values of initial macromolecular concentration, $C_{10} + C_{20}$, shown. (———) Gradient curve; (- - -) pHA. The relative area of the slow-moving peak is given in parentheses. Times of electrophoresis: $C_{10} + C_{20} = 8.75 \times 10^{-5}$ M, 7.17×10^3 sec; all other values of $C_{10} + C_{20}$, 6.92×10^3 sec. pHA scale: 2.00–2.02. [From Cann and Goad (30).]

reproduce very closely the observed dependence of the patterns of serum albumin in 0.01 M acetate buffer, pH 4, upon protein concentration [see Cann (34, Fig. 1)]. Thus, the relative area of the slow-moving peak decreases progressively as the macromolecular concentration is lowered. Concomitantly, the migration velocity of the fast peak decreases, approaching as a limit the weighted average of the electrophoteric mobilities of uncomplexed and complexed protein in the initial solution. This behavior is understandable in terms of the progressively decreasing strength of the gradients of HA which determine resolution of the reaction boundary into two peaks. At sufficiently low macromolecular concentration, the concentration of HA is in such great excess as to be considered constant. Under such conditions, the initial equilibrium composition is maintained during electrophoretic transport. Consequently, the patterns show a single peak migrating with the weighted average mobility.

This is the theoretical basis for one of the precautions (*39*) required to assure validity of fractionation as a means of determining whether multiple electrophoretic peaks are indicative of inherent heterogeneity or interaction. In the fractionation test, the protein disappearing across the faster-moving ascending and slower-moving descending peaks is isolated. The resulting fractions are then analyzed electrophoretically under conditions identical with those used in the original separation. For interactions, the fractions will behave like the unfractionated material and show multiple peaks, while for heterogeneity, a single peak will be obtained. Consider, for example, the patterns shown in Fig. 72 for $n = 30$. If, in practice, the protein disappearing across the faster-moving ascending peak were to be removed from the electrophoresis cell, dialyzed against buffer, and subjected to electrophoresis *without restoring the protein concentration used in the original separation*, the ascending as well as the descending pattern of the fraction might show a single peak, thereby leading to an incorrect conclusion of heterogeneity. It is imperative, therefore, either to concentrate the fraction before dialysis or to calibrate the method by analyzing the unfractionated material at several different concentrations. Reequilibration of the fraction by dialysis against buffer prior to analysis is an absolute requirement[2]; otherwise, a fraction from an interacting system will show a single peak, once again leading to an incorrect conclusion of heterogeneity [see Cann (*35*, Fig. 2, control in Fig. 6 under the heading Continuous Electrophoresis B)].

As already mentioned, the computed patterns predict the convective disturbances observed in the ascending patterns of ovalbumin and serum albumin in acetic acid-containing media at pH 4 (*37*). This is illustrated in Figs. 74 and 75. After about 7×10^3 sec of electrophoresis, the theoretical ascending patterns show a slight elevation of the protein gradient above the baseline between the two well-resolved peaks. This plateau is maintained until about 10^4 sec. For longer times of electrophoresis, the gradient between

---

[2] A fraction could also be restored to the original salt concentration and pH by addition of a small quantity of a buffer of appropriate composition (*35*), but this procedure is awkward.

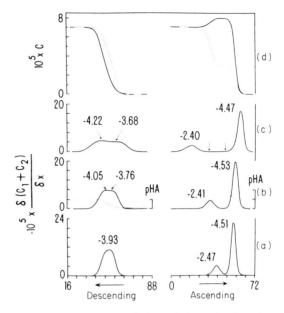

FIG. 74. Time course of development of theoretical moving-boundary electrophoretic patterns computed for $\mu_1 = -5 \times 10^{-5}$ cm^2 sec^{-1} V^{-1}, $\mu_2 = -3 \times 10^{-5}$, $K^{1/n} = 10^2$, $C_{20}/C_{10} = 1$, and $n = 30$. Three lower graphs: (———) gradient curve; (– – –) pHA. Topmost graphs: (———) C_1; (– – –) C_2. Time of electrophoresis: (a) 6.92×10^3 sec; (b) and (d), 1.04×10^4 sec; (c) 2.01×10^4 sec. pHA scale: ascending, 1.98–2.00; descending, 2.00–2.02. Note the effect of reequilibration on the concentrations of P and P(HA)$_n$ during differential transport, as illustrated in (d). [From Cann and Goad (42).]

the peaks changes sign and the curve becomes slightly depressed below the base line in the region bounded by the *vertical arrows* in the patterns. In other words, the electrophoretic process generates inverted gradients of protein concentration between the peaks. Consequently, the density of the protein solution increases with increasing height in this region of the electrophoresis column. Clearly, such a situation would be an unstable one in free moving-boundary electrophoresis and would result in convective mixing. This is exactly what happens experimentally. For early times of electrophoresis, there is a gradient of protein concentration between the two ascending peaks. Eventually, however, this gradient is obliterated by mild convection. Very small convective pips sometimes develop along the trailing edge of the fast peak and the leading edge of the slow one. In two instances, highly localized, inverted refractive index gradients have been recorded. A possible explanation is that a localized, inverted gradient of protein concentration is, at least momentarily, stabilized by an opposing gradient of salt concentration. This would, of course, be a rare event.

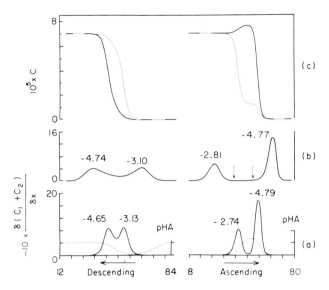

FIG. 75. Time course of development of theoretical moving-boundary electrophoretic patterns computed for same values of the parameters as in Fig. 74 except for $n = 60$. Time of electrophoresis: (a) and (c), 6.92×10^3 sec; (b) 2.01×10^4 sec. [From Cann and Goad (42).]

The discussion so far has dealt with protein–small-molecule interactions for situations in which the protein is acid to its isoelectric point (positive mobilities and mobility of the complexed protein molecule assumed to be greater than that of the uncomplexed protein). However, electrophoretic patterns have also been computed for the other case, in which the protein is alkaline to its isoelectric point (negative mobilities and mobility of complexed protein smaller than uncomplexed protein). The results are displayed in Figs. 74 and 75. Except for negative migration velocities, these patterns are virtually indistinguishable from those computed for a protein on the acid side of its isoelectric point (compare these patterns with those for comparable values of n in Fig. 72). There are at least two known cases of protein–buffer interaction at pH values alkaline to the isoelectric point. One of these (41) is BSA in phosphate–borate buffer, pH 6.2, or borate buffer, pH 8.9; the other is β-lactoglobulin in media containing acetate buffer (private communication from Dr. Serge N. Timasheff). β-Lactoglobulin is a particularly interesting case, since it differs from most other proteins in that its electrophoretic behavior reveals protein–acetate buffer interaction at pH values both alkaline and acid to its isoelectric point. The patterns exhibit the characteristic nonenantiography associated with such interaction and are sensitive to buffer concentration. Fractionation experiments have established that the two peaks in the patterns arise from interaction, not from inherent heterogeneity.

From the foregoing discussion, it is apparent that theory adequately accounts for observed electrophoretic behavior of proteins in carboxylic acid-containing media. However, the computations have other, even broader implications. The most important of these for the biologist using electrophoresis ancillary to his research is illustrated in Fig. 72. The computed patterns presented in this figure show that the whole spectrum of experimentally recognized types of moving-boundary patterns may, in principle, arise from interactions of a single macromolecule with an uncharged constituent of the solvent medium even when instantaneous establishment of equilibrium is assumed. In particular, if the interaction is very highly cooperative, the system will migrate like a mixture of two stable, noninteracting macromolecules. (See, for example, $n = 300$; but bear in mind that similar patterns will be obtained at much smaller n if the concentration of small molecule is lower.) In other words, the patterns will be enantiographic and will show two well-resolved peaks. Such patterns might, in practice, be misinterpreted if electrophoretic fractionation and tests for biological activity are not accompanied by electrophoretic analyses of the fractions under conditions identical with those used in the original separation. It is also important to note in passing that the small, neutral molecule is not restricted to undissociated buffer acids. Neutral buffer bases and deliberately added small, neutral reactants would also be effective in producing these effects. It is precisely this consideration that suggests new applications of electrophoresis to biological problems. For example, the electrophoresis of a purified antiuncharged hapten antibody in a solvent containing the appropriate concentration of the hapten, e.g., p-amino-β-phenyllactoside, might yield important information concerning the mechanism of the antigen–antibody reaction. Thus, if the antibody–uncharged hapten complex has a different electrophoretic mobility than the uncombined antibody molecule, the electrophoretic patterns will show at least two peaks. Similar experiments might be performed with enzymes possessing appropriate association constants for their uncharged substrates or competitive inhibitors.

Results for Zone Electrophoresis. Upon realization that a single macromolecule can give enantiographic moving-boundary patterns showing two peaks, the computations were immediately extended (*40*) to include zone electrophoresis on a solid support or in a density gradient. As with moving-boundary electrophoresis, the nature of the theoretical zone electrophoretic patterns is determined by the interplay of three factors: (1) the value of n, and (2) the concentration gradients of HA produced by (3) the continual reequilibration of reaction (261) during differential transport of P and $P(HA)_n$. For relatively small values of n, the patterns show a single, slightly skewed zone (Fig. 76).

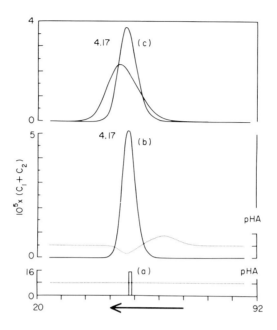

FIG. 76. Time course of development of theoretical zone-electrophoretic pattern for either of the following two sets of conditions: $K^{1/n} = 10^4$, $C_{20}/C_{10} = 1$, $n = 3$, pHA scale is 3.9–4.1; or $K^{1/n} = 10^2$, $C_{20}/C_{10} = 1$, $n = 30$, pHA scale is 1.99–2.01. (———) Electrophoretic pattern; (- - -) pHA. Time of electrophoresis: (a) $t = 0$; (b) 3.46×10^3 sec; (c) 6.92×10^3 sec; (d) 2.01×10^4 sec. The abscissa in this and all other figures displaying theoretical zone patterns is $10x + 51.5 + 4 \times 10^{-3}t$. [From Cann and Goad (40).]

For intermediate values (Fig. 77), the patterns resolve into two zones for short times of electrophoresis, although the relative amounts of macromolecule contained in the zones do not correspond to the initial equilibrium composition of the macromolecule solution. As electrophoresis proceeds, the slower-moving, minor zone spreads excessively as the two zones blend into a single trailing one which eventually becomes very broad indeed. Thus, for the longest time of electrophoresis (Fig. 77f), the breadth of the zone is about 7 cm, the maximum having migrated 9.3 cm. Such broad trailing zones are often encountered in practice, and are usually attributed either to adsorption of the protein on the solid support, as with serum albumin on paper, or to known inherent heterogeneity, as in the case of γ-globulin on starch. Clearly, in the absence of demonstrated heterogeneity, interaction of a single macromolecule with an uncharged constituent of the solvent medium must also be entertained as a possible explanation of trailing zones.

The results obtained for large values of n (Fig. 78) are of tremendous interest, since these patterns show two stable zones. The relative amounts of macromolecule in the two zones approach the initial equilibrium composition, but, as is typical of reaction patterns, the concentration never reaches zero between the zones. Thus, a single macromolecule which interacts cooperatively and reversibly with an uncharged constituent of the solvent, with concomitant change in electrophoretic mobility, can give two well-resolved and intense zones despite instantaneous establishment of equilibrium. The small, uncharged constituent may be undissociated buffer acid, buffer base, a contaminant, or intentionally added reactant.

This important prediction has been verified experimentally (*41*) by zone electrophoresis of BSA on cellulose acetate. Electrophoresis was carried out in either phosphate buffer, pH 6.1 and ionic strength 0.043, or a phosphate–borate buffer of about the same pH and ionic strength. The zone patterns are

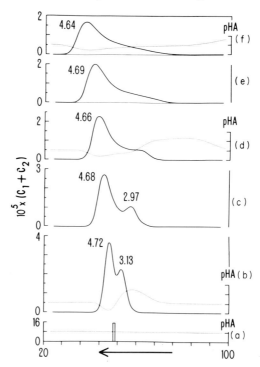

FIG. 77. Time course of development of theoretical zone-electrophoretic pattern for either of the following two sets of conditions: $K^{1/n} = 10^4$, $C_{20}/C_{10} = 1$, $n = 10$, pHA scale in (b) and (d) is 3.9–4.2 and in (f) is 3.9–4.1; or $K^{1/n} = 10^2$, $C_{20}/C_{10} = 1$, $n = 100$, pHA scale in (b) and (d) is 1.99–2.02 and in (f) is 1.99–2.01. (———) Electrophoretic pattern; (– – –) pHA. Time of electrophoresis: (a) $t = 0$; (b) 3.46×10^3 sec; (c) 6.92×10^3 sec; (d) 1.04×10^4 sec; (e) 1.38×10^4 sec; (f) 2.01×10^4 sec. [From Cann and Goad (*40*).]

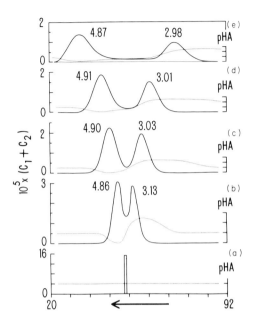

FIG. 78. Time course of development of theoretical zone-electrophoretic pattern for either of the following two sets of conditions: $K^{1/n} = 10^4$, $C_{20}/C_{10} = 1$, $n = 30$, pHA scale is 3.95–4.10; or $K^{1/n} = 10^2$, $C_{20}/C_{10} = 1$, $n = 300$, pHA scale is 1.995–2.010. (——) Electrophoretic pattern; (– – –) pHA. Time of electrophoresis: (a) $t = 0$; (b) 3.46×10^3 sec; (c) 6.92×10^3 sec; (d) 1.04×10^4 sec; (e) 2.01×10^4 sec. [From Cann and Goad (40).]

presented in Figs. 79a and b; while the protein gave a single zone in phosphate buffer, two zones were obtained in phosphate–borate, the faster-moving zone being the major one. In order to determine whether the two zones reflect interaction or heterogeneity, the protein was eluted from each of the unstained zones with phosphate–borate buffer and the resulting fractions were analyzed by zone electrophoresis in the same buffer. Like unfractionated BSA, the pattern of each fraction showed two zones (Figs. 79c and d), which demonstrates that the duplicity of zones arises from reversible interaction of BSA with boric acid or borate anion and is not due to inherent heterogeneity. In the case of the fractions, which were analyzed at rather low protein concentrations, the slower of the two zones was the major one. It is particularly interesting that such a large proportion of the fast fraction migrated in the slower zone when reanalyzed (Fig. 79d). This result underscores the conclusion that the two zones arise from interaction.

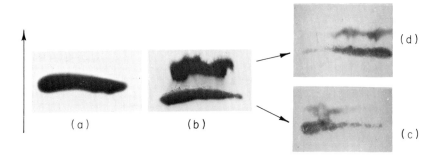

FIG. 79. Demonstration that the two zones shown by BSA in phosphate–borate buffer on cellulose acetate arise from reversible interaction of the protein with boric acid or borate anion and are not due to an inherent heterogeneity: (a) zone pattern of BSA in 0.0043 M Na_2HPO_4–0.03 M NaH_2PO_4, pH 6.07, 5% protein; (b) pattern in 0.012 M Na_2HPO_4–0.0081 M NaH_2PO_4–0.4 M H_3BO_3, pH 6.15, 5% (each of the two zones obtained in the phosphate–borate buffer was cut from unstained strips and the protein eluted with buffer, and the resulting fractions were then analyzed in the same phosphate–borate buffer); (c) zone pattern of material eluted from slower-migrating zone; (d) pattern of material from faster zone. Solid vertical arrow indicates direction of migration. [From Cann (41).]

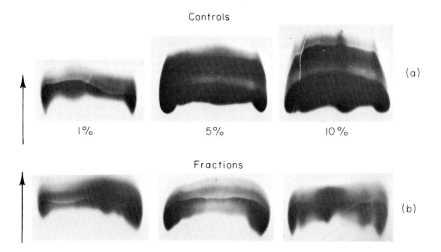

FIG. 80. Demonstration that the two or three zones shown by BSA in borate buffer (0.0108 M NaOH–0.260 M H_3BO_3 originally added to prepare buffer, pH 7.15) on cellulose acetate arise from interaction of the protein with the buffer and are not due to an inherent heterogeneity. Zone patterns of (a) unfractionated controls at three different protein concentrations (0.5% protein showed only a single zone; the fastest zone shown by 10% protein shaded off in the direction of migration, but the shaded portion could not be captured photographically), (b) three representative fractions obtained from 10% protein by cutting the slowest zone from unstrained strips and removing the protein by centrifugation. Vertical arrow indicates direction of migration. [From Cann (41).]

Zone electrophoresis of BSA has also been carried out in sodium borate buffer at pH 7.2 and ionic strength 0.011 (Fig. 80). This system exhibits the interesting property that, although resolution into two or three zones was excellent at protein concentrations in the range 1–10%, patterns obtained at lower concentrations showed only a single zone (5% BSA in 0.011 M phosphate buffer, pH 7.1, gave a single zone). Accordingly, fractionation experiments to decide between interaction and heterogeneity were carried out at 10% protein by removing the material from the slowest-moving (major) zone. This was done in order to obtain fractions of sufficiently high concentration to ensure against possible misinterpretation in the event that electrophoresis of the fractions should give a single zone. In fact, a preliminary experiment at 5% protein gave a fraction whose concentration was inadvertently so low that its electrophoretic pattern showed only a single zone, as in a control analysis of unfractionated protein at low concentration. On the other hand, the patterns of fractions obtained at 10% protein consistently showed, not one, but two or three zones. It is apparent, therefore, that the multiple zones arise from interaction of BSA with borate buffer and are not due to inherent heterogeneity.

The phenomenon of multiple zones due to reversible protein–buffer interaction is not unique for BSA. Thus, conalbumin exhibits one, two, or three zones on starch gel electrophoresis at pH 8.9, depending systematically upon the protein concentration (*43*). It has been demonstrated that the multiple zones shown at intermediate and high concentrations arise from reversible interaction of the protein with borate buffer. Also, human myoglobin gives three zones on starch gel in a Tris–borate buffer due to a reaction (*44*), although it is not clear whether it is a protein–buffer interaction. In all these cases, aggregation reactions have been eliminated. It appears likely that multiplicity of zones due to interaction may be of more general occurrence than recognized. Furthermore, it is a disturbing fact that such patterns could easily be misinterpreted as indicating inherent heterogeneity. Fortunately, however, with certain precautions, fractionation provides an unambiguous method for distinguishing between interaction and heterogeneity. The various measures which must be taken to ensure correct interpretation of electrophoretic patterns will be detailed in Chapter VI.

Finally, these various electrophoretic considerations should also apply to analytical and zone sedimentation, with the usual proviso of pressure insensitivity. It is conceivable, for example, that cooperative binding of small molecules or ions by a protein or other large molecule could cause a significant change in its sedimentation coefficient due to alteration in macromolecular conformation. A small change in sedimentation coefficient has, in fact, been observed for the interaction of aspartate transcarbamylase with its ligands (see Chapter II). Given the appropriate conditions, such a system might show a bimodal reaction zone upon sedimentation through a density gradient containing the ligand. Still another possibility not encompassed by the foregoing theory is that binding of ligand favors molecular association or dissociation of the protein. The consequences of this class of interactions are explored below.

Extension of Theory to Ligand-Mediated Association–Dissociation Reactions. The foregoing theory has been extended (42, 45) to include velocity sedimentation of systems of molecules that interact reversibly according to the reaction

$$m\mathrm{M} + n\mathrm{X} \rightleftharpoons \mathrm{M}_m\mathrm{X}_n \tag{265}$$

in which a macromolecule M associates into an m-mer with the mediation of a small molecule or ion X of which a fixed number n are bound into the complex. We refer to this class of interactions as *ligand-mediated association* to distinguish it from simple association–dissociation reactions of the type $m\mathrm{M} \rightleftharpoons \mathrm{M}_m$. One can visualize several mechanisms whereby the binding of small ligand molecules favors association of the protein. For example, bivalent metal ions might bond monomeric units together simply by forming bridges between carboxylate, imidazole, or sulfhydryl groups without causing significant alteration in macromolecular conformation. On the other hand, it might be that the monomer conformation that favors formation of the m-mer is stabilized by binding of a number of ligand molecules, which would constitute one kind of allosteric interaction. In this light, the reaction in which binding of small molecules causes dissociation of a macromolecule into subunits is also of interest.

The set of conservation equations for sedimentation of ligand-mediated associating systems in a sector-shaped cell takes the form

$$\frac{\partial(C_1 + mC_2)}{\partial t} = \frac{1}{r}\frac{\partial}{\partial r}\left[\left(D_1 \frac{\partial C_1}{\partial r} - C_1 s_1 \omega^2 r\right)r + m\left(D_2 \frac{\partial C_2}{\partial r} - C_2 s_2 \omega^2 r\right)r\right]$$

$$\frac{\partial(nC_2 + C_3)}{\partial t} = \frac{1}{r}\frac{\partial}{\partial r}\left[n\left(D_2 \frac{\partial C_2}{\partial r} - C_2 s_2 \omega^2 r\right)r + \left(D_3 \frac{\partial C_3}{\partial r} - C_3 s_3 \omega^2 r\right)r\right]$$

(266)

where the subscripts 1, 2, and 3 designate M, $\mathrm{M}_m\mathrm{X}_n$, and X, respectively. The calculations have been made for rates of reaction very much larger than the rates of diffusion and sedimentation, so that local equilibrium attains at every instant. In this limit,

$$C_2 = KC_1^m C_3^n \tag{267}$$

where K is the equilibrium constant of the reaction. These equations have been solved numerically (as described in Chapter V) for both dimerization and tetramerization reactions, and representative theoretical sedimentation patterns are displayed below as plots of $\delta(C_1 + mC_2)/\delta r$ vs. r.

In certain respects, the results for dimerization are the more revealing with regard to fundamental principles. While the Gilbert theory correctly predicts that the sedimentation pattern of the system $2M \rightleftharpoons M_2$ will show only a single peak for instantaneous establishment of equilibrium, one might expect *ligand-mediated dimerization* to show two peaks due to production of concentration gradients of the unbound small molecule (or ion) in the sedimentation column. This would be so even for instantaneous reestablishment of equilibrium during differential transport of monomer and dimer, at least for high centrifugal fields. At sufficiently lower fields with correspondingly longer times of centrifugation, sufficiently strong gradients of the small molecule could not be maintained against diffusion and, consequently, resolution of the reaction boundary into two peaks would not occur. In the limit where, for any reason, the concentration of small molecule in the sedimentation column is not significantly perturbed by the reaction during differential sedimentation of the macromolecular species, the system effectively approaches the case of simple dimerization considered by Gilbert. The theoretical calculations predict precisely the behavior.

The patterns shown in Fig. 81 were calculated using values of macromolecular concentration and parameters chosen so as to approximate the sedimentation of high molecular weight DNA (about 10^8 Daltons) at three different values of the centrifugal field, i.e., at three different rotor speeds. Dimerization was considered to be brought about as the result of reversible binding into the complex of 30 molecules of the small molecule (or ion) present at a constituent concentration 30 times that of the macromolecule. As anticipated above, the pattern computed for the highest rotor speed shows two peaks—a broad, rapidly sedimenting peak having a sedimentation coefficient slightly less than that of the dimer and a sharp, slow peak with the sedimentation coefficient of the monomer. The slow peak is the minor one, but as the centrifugal field is lowered, it grows at the expense of the fast peak. At the lowest rotor speed, the patterns show essentially a single peak having a sedimentation coefficient about 30% greater than that of the monomer, with only a vestige of the fast peak. These calculations were stimulated by the experimentally observed (46–48) sedimentation behavior of high-molecular-weight DNA, which has been interpreted (47) in terms of a phase change dependent upon rotor speed and involving kinetically controlled association. A conclusion to be drawn from the results of these and previously described theoretical calculations is that there are three possible mechanisms which can give rise to rotor-speed-dependent sedimentation behavior: (1) kinetically controlled interactions; (2) pressure-sensitive equilibria; and now (3) rapidly equilibrating ligand-mediated association.

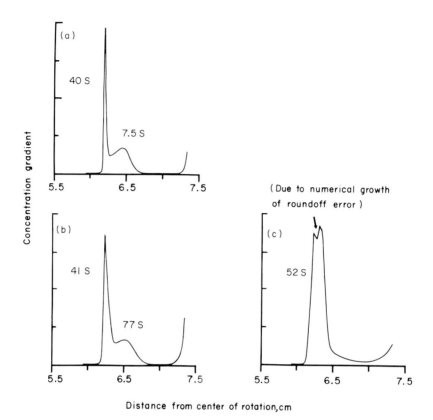

FIG. 81. Theoretical sedimentation patterns computed for the ligand-mediated dimerization reaction $2M + 30X \rightleftharpoons M_2X_{30}$ at three different values of the centrifugal field. Rotor speed and time of sedimentation shown under each pattern. Ordinate scales: (a) 60,000 rpm, 3×10^2 sec, 0 to 2×10^{-9} mole liter^{-1} cm^{-1}; (b) 20,000 rpm, 3×10^3 sec, 0 to 1.5×10^{-9} mole liter^{-1} cm^{-1}; (c) 6000 rpm, 3×10^4 sec, 0 to 8×10^{-10} mole liter^{-1} cm^{-1}. The following values of the macromolecular concentrations and parameters were chosen to approximate the sedimentation of high molecular weight DNA: $C_{10} = 10^{-10}$ M, $C_{20} = 5 \times 10^{-11}$ M, $s_1 = 40$ S, $s_2 = 80$ S, $D_1 = D_2 = 10^{-9}$ cm^2 sec^{-1}. The small-molecule concentration and other parameters were assigned the values: $C_{30} = 4.5 \times 10^{-9}$ M, $s_3 = 0.1$ S, $D_3 = 10^{-6}$ cm^2 sec^{-1}. [From Cann and Goad (42).]

Computations have also been made for a protein of molecular weight about 60,000. The theoretical pattern presented in Fig. 82 is for dimerization with the mediation of 30 small molecules. For the particular choice of small-molecule concentration used in this calculation, the reaction boundary shows two well-resolved peaks with sedimentation coefficients equal to those of the monomer and dimer, respectively. The exceedingly high resolution reflects

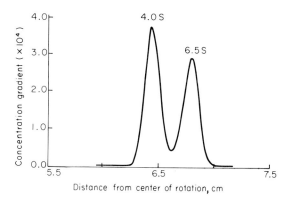

FIG. 82. Theoretical sedimentation pattern computed for the ligand-mediated dimerization $2M + 30X \rightleftharpoons M_2X_{30}$. The following values of the macromolecular concentrations and parameters were chosen to approximate the sedimentation of a protein of molecular weight about 60,000: $C_{10} = 7 \times 10^{-5}$ M, $C_{20} = 3.5 \times 10^{-5}$ M, $s_1 = 4$ S, $s_2 = 6.35$ S, $D_1 = 6 \times 10^{-7}$ cm² sec⁻¹, $D_2 = 4.76 \times 10^{-7}$ cm² sec⁻¹. The small-molecule concentration and other parameters were assigned the values: $C_{30} = 10^{-5}$ M, $s_3 = 0.1$ S, $D_3 = 10^{-5}$ cm² sec⁻¹. Rotor speed is 60,000 rpm; time of sedimentation is 5540 sec. [From Goad and Cann (45).]

the cooperativeness of the interaction. Upon increasing the unbound small-molecule concentration progressively from 10^{-5} to 1 M, at the same time holding the degree of dimerization constant, the slow peak grows steadily at the expense of the fast one, with concomitant drift in sedimentation velocity toward the weight-average value until resolution disappears entirely at the highest concentration. In the limit of sufficiently high concentration of the small molecule (or low centrifugal field), the Gilbert theory is valid because the concentration of small molecule cannot be perturbed significantly under these conditions. Recently, we have made calculations which reveal a similar behavior for dimerization mediated by the binding of only a single small molecule into the complex. The patterns for 50% dimerization displayed in Fig. 83 illustrate the way in which resolution decreases with increasing ligand concentration as the Gilbert limit is approached. Analyses of the type presented in Fig. 84 have established that resolution is dependent upon the production of small-molecule concentration gradients by rapid reequilibration during differential transport of monomer and dimer. When such gradients are sufficiently strong that the macromolecular composition also changes markedly along the centrifuge cell, resolution will occur. Thus, in the example quoted, the perturbation of the small-molecule concentration is so severe that the slowly sedimenting peak in the reaction boundary corresponds to an equilibrium mixture containing more than 90% monomer, while the region of the faster peak contains

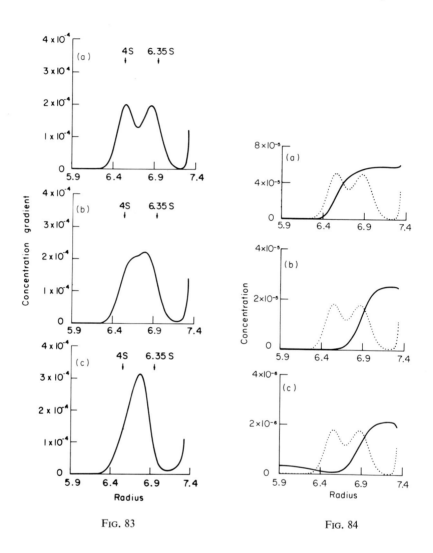

FIG. 83. Theoretical sedimentation patterns for the ligand-mediated dimerization reaction $2M + X \rightleftharpoons M_2X$. Dependence of boundary shape at 50% dimerization upon the initial concentration of unbound small ligand: (a) $C_{30} = 2 \times 10^{-6}$ M; (b) 5×10^{-6} M; (c) 3.5×10^{-5} M. Other parameters are the same as those in Fig. 82 except for the time of sedimentation, which is 6431.4 sec here.

FIG. 84. Integral curves corresponding to the total macromolecular gradient curve shown in Fig. 83a: plot of concentrations of (a) monomer, (b) dimer, and (c) unbound small ligand molecule versus radius. The total macromolecular gradient curve (dotted) is reproduced in each panel to permit comparison.

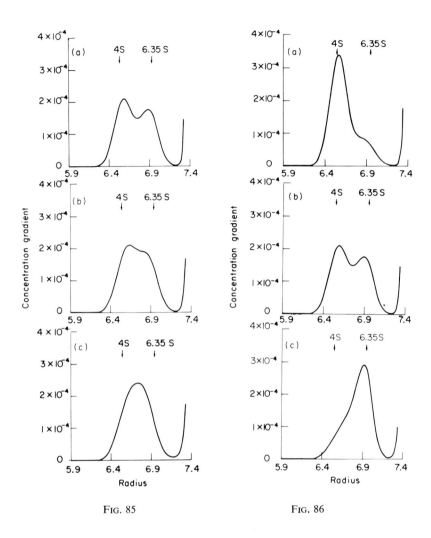

FIG. 85.   FIG. 86.

FIG. 85. Theoretical sedimentation patterns for the ligand-mediated dimerization reaction $2M + 6X \rightleftharpoons M_2X_6$. Dependence of boundary shape at 50% dimerization upon the initial concentration of unbound small ligand: (a) $C_{30} = 5 \times 10^{-5}\ M$; (b) $10^{-4}\ M$; (c) $2 \times 10^{-4}\ M$. Other parameters as in Fig. 82 except for the time of sedimentation, which is 6431.4 sec here.

FIG. 86. Theoretical sedimentation patterns for the ligand-mediated dimerization reaction $2M + 6X \rightleftharpoons M_2X_6$. Dependence of boundary shape upon degree of dimerization, $K = 4.57 \times 10^{29} M^{-7}$: (a) 25% dimerization, $C_{30} = 3.89 \times 10^{-5}\ M$; (b) 50%, $5 \times 10^{-5}\ M$; (c) 75%, $6.74 \times 10^{-5}\ M$. $C_{10} + 2C_{20} = 14 \times 10^{-5}\ M$; time of sedimentation is 6431.4 sec; other parameters as in Fig. 82.

up to 50% dimer. Clearly, such nonlinear coupling of transport with macromolecule–small-molecule interaction will be sensitive to the cooperativeness of the interaction. Consider, for example, the patterns computed for dimerization mediated by the binding of six small molecules into the complex (Fig. 85). Comparison with the analogous patterns for mediation by a single small molecule shows that resolution can occur at higher ligand concentrations when the interaction is cooperative. In general, the higher the cooperativeness, the wider is the range of ligand concentrations over which resolution can occur. Finally, the set of patterns presented in Fig. 86 for mediation by six ligand molecules illustrates how the shape of the reaction boundary depends upon the initial degree of dimerization as determined by total ligand concentration at constant macromolecule concentration and fixed equilibrium constant. Although the fast peak grows at the expense of the slow one as the degree of dimerization increases, we note that the areas under the peaks do not faithfully reflect the initial composition; nor do the peaks sediment with the sedimentation coefficients of monomer and dimer. Under some conditions, the slow peak sediments more rapidly than the monomer since it contains dimer, while the fast peak always sediments slower than the dimer since it contains monomer.

Tetramerization reactions have also been considered. The theoretical sedimentation patterns for the *ligand-mediated tetramerization* $4M + 4X \rightleftharpoons M_4X_4$ show two peaks whose proportions vary systematically with macromolecular composition (Fig. 87). Thus, the fast peak grows at the expense of the slow one as the degree of tetramerization is increased by increasing the small-molecule concentration at constant macromolecular concentration and fixed equilibrium constant. Each pattern corresponds to an initial equilibrium mixture whose composition is defined by an apparent association constant $KC_{30}^4 = C_{20}/C_{10}^4$; and in principle, one can seek limiting values of C_{30} and K such that the shape of the boundary approaches the shape predicted by the Gilbert theory for simple tetramerization, $4M \rightleftharpoons M_4$. Since the values of C_{30} used in our calculations are less than the limiting ones, the concentration of small molecules is significantly perturbed by reequilibration during differential transport of monomer and tetramer. Consequently, the patterns deviate significantly from Gilbert patterns. In particular, the area of the slow peak (as judged from the total macromolecule concentration Δ_s corresponding to the minimum in the gradient curve) is 20–50% smaller than predicted by the Gilbert theory as elaborated by Fujita (49) for a sector-shaped cell. In practice, this observation precludes calculation of the apparent association constant from Δ_s, which is presumably a valid procedure for simple tetramerization. Even if Fujita's relationship between K', Δ_s, and n were to be employed instead of Eq. (216) of Chapter III, the calculated value of KC_{30}^4 could be in error by as much as a factor of 7.

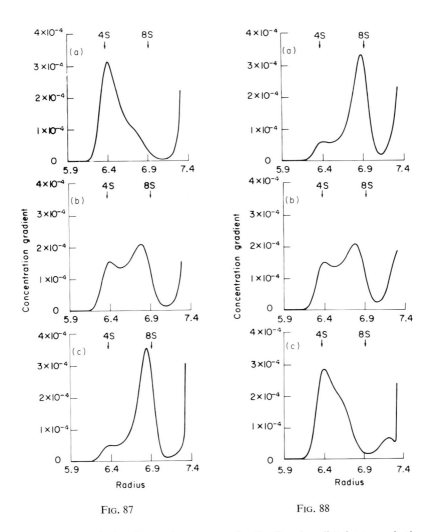

FIG. 87. Theoretical sedimentation patterns for the ligand-mediated tetramerization reaction $4M + 4X \rightleftharpoons M_4X_4$. Dependence of boundary shape upon degree of tetramerization, $K = 3.04 \times 10^{28} M^{-7}$: (a) 25% tetramerization, $C_{30} = 3.924 \times 10^{-5} M$; (b) 50%, $7 \times 10^{-5} M$; (c) 75%, $1.549 \times 10^{-4} M$. $C_{10} + 4C_{20} = 14 \times 10^{-5} M$; $s_2 = 8$ S; $D_2 = 3.8 \times 10^{-7}$ cm² sec⁻¹; time of sedimentation is 5046 sec; other parameters as in Fig. 82.

Fig. 88. Theoretical sedimentation patterns for the ligand-mediated dissociation of a tetrameric macromolecule into its subunits, $M_4 + 4X \rightleftharpoons 4MX$. Dependence of boundary shape upon degree of dissociation, $K_d = 5.71 \times 10^4 M^{-1}$: (a) 25% dissociation, $C_{30} = 3.163 \times 10^{-5} M$; (b) 50%, $7 \times 10^{-5} M$; (c) 75%, $1.249 \times 10^{-4} M$. Time of sedimentation is 5146.1 sec; other parameters as in Fig. 87.

Because *ligand-mediated dissociation* reactions are of considerable interest for biological regulatory mechanisms, calculations have been made for a tetrameric protein which dissociates into its identical subunits as the result of reversible binding by each subunit of a single ligand molecule, $M_4 + 4X \rightleftharpoons 4MX$. The computed sedimentation patterns (Fig. 88) are in many respects strikingly similar to those for ligand mediated tetramerization, and the same preclusion applies to the calculation of the apparent equilibrium constant. A unique feature of the patterns, however, is the subsidiary peak which develops with time near the bottom of the cell. Although the peak is evident only in Fig. 88c, it develops in the other two patterns at later times of sedimentation. It is apparent from the integral plots presented in Fig. 89 that the

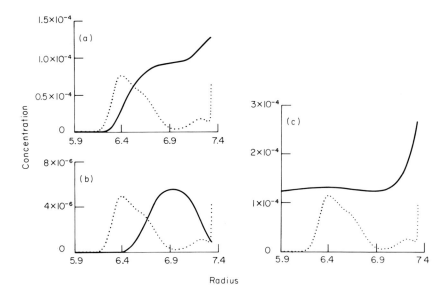

FIG. 89. Integral curves corresponding to the total macromolecular gradient curve shown in Fig. 88c: plot of concentrations of (a) monomer, (b) tetramer, and (c) unbound small ligand molecule versus radius. The total macromolecular gradient curve (dotted) is reproduced in each panel to permit comparison.

following molecular mechanism is responsible for generation of the subsidiary peak: As the macromolecules accumulate at the bottom of the centrifuge cell, reequilibration releases transported ligand molecules, which then diffuse back toward the top, thereby increasing the ligand concentration in the region extending several millimeters from the bottom. This, in turn, causes dissociation of sedimenting tetramer molecules into slower-moving subunits which tend to accumulate in said region.

In conclusion, these results hold promise of wide application for the study of associating–dissociating systems. For many systems, there is either direct evidence for believing that binding of small molecules or ions causes association of the protein, or circumstantial evidence, such as an effect of rotor speed upon the sedimentation patterns. Some examples follow: enzymes such as *Bacillus subtilis* α-amylase, which is apparently composed of two subunits held together by a single Zn^{2+} (*50*); aggregation of a variety of proteins by Al^{3+} (*51*); and the effect of anion binding on aggregation of proteins such as γ-globulin at acid pH (*52*). Lamprey hemoglobin is an example of a protein which dissociates into subunits upon binding of its specific ligand, oxygen (*53*). Clearly, quantitative interpretation of the sedimentation patterns of such systems depends upon knowledge as to the rate of equilibration. Only when the time required to establish the monomer–polymer equilibrium is long in comparison with the time of the sedimentation experiment can the areas and sedimentation coefficients of the peaks in the pattern be interpreted in the classical fashion. An exemplary investigation is the study by Klapper and co-workers (*54, 55*) of the thiocyanate-promoted dissociation of hemerythrin into its monomeric subunits.

REFERENCES

1. E. O. Field and A. G. Ogston, *Biochem. J.* **60**, 661 (1955).
2. L. W. Nichol, J. L. Bethune, G. Kegeles, and E. L. Hess, *in* "The Proteins" (H. Neurath, ed.), Vol. II, 2nd ed., Chapter 9, p. 352. Academic Press, New York, 1964.
3. J. L. Bethune and G. Kegeles, *J. Phys. Chem.* **65**, 1761 (1961).
4. M. S. N. Rao, and G. Kegeles, *J. Am. Chem. Soc.* **80**, 5724 (1958).
5. D. F. Oberhauser, J. L. Bethune, and G. Kegeles, *Biochemistry* **4**, 1878 (1965).
6. G. G. Belford and R. L. Belford, *J. Chem. Phys.* **37**, 1926 (1962).
7. J. R. Cann and H. R. Bailey, *Arch. Biochem. Biophys.* **93**, 576 (1961).
8. J. C. Giddings, *J. Chromatog.* **3**, 443 (1960).
9. R. A. Keeler and J. C. Giddings, *J. Chromatog.* **3**, 205 (1960).
10. A. Klinkenberg, *in* "Gas Chromatography, 1960" (R. P. W. Scott, ed.), p. 386. Butterworths, Washington, 1960.
11. A. Klinkenberg, *Chem. Eng. Sci.* **15**, 255 (1961).
12. P. C. Scholten, *Arch. Biochem. Biophys.* **93**, 568 (1961).
13. G. Kegeles, L. Rhodes, and J. L. Bethune, *Proc. Natl. Acad. Sci.* **58**, 45 (1967).
14. R. Josephs and W. F. Harrington, *Biochemistry* **5**, 3474 (1966).
15. R. Josephs and W. F. Harrington, *Proc. Natl. Acad. Sci.* **58**, 1587 (1967).
16. H. K. Schachman, "Ultracentrifugation in Biochemistry," pp. 60–62. Academic Press, New York, 1959.
17. H. K. Schachman and W. F. Harrington, *J. Polymer Sci.* **12**, 379 (1954).
18. J. L. Bethune and G. Kegeles, *J. Phys. Chem.* **65**, 433 (1961).
19. J. L. Bethune and G. Kegeles, *J. Phys. Chem.* **65**, 1755 (1961).
20. L. C. Craig and D. Craig, *in* "Techniques of Organic Chemistry" (A. Weissberger, ed.), Vol. III. Wiley (Interscience), New York, 1950.
21. I. P. Crawford and C. Yanofsky, *Proc. Natl. Acad. Sci.* **44**, 1161 (1958).

22. L. F. Ten Eyck and W. Kauzmann, *Proc. Natl. Acad. Sci.* **58,** 888 (1967).
23. J. Monod, J. Wyman, and J.-P. Changeux, *J. Mol. Biol.* **12,** 88 (1965).
24. G. B. Kitto, P. M. Wasserman, and N. O. Kaplan, *Proc. Natl. Acad. Sci.* **56,** 578 (1966).
25. J. R. S. Fincham and A. Coddington, *Symp. Quant. Biol. Cold Spring Harbor* **28,** 517 (1963).
26. J. R. S. Fincham, "Genetic Complementation." Benjamin, New York, 1966 (see particularly pp. 80–89).
27. J. R. Sadler and A. Novick, *J. Mol. Biol.* **12,** 305 (1965).
28. W. Gilbert and B. Müller-Hill, *Proc. Natl. Acad. Sci.* **56,** 1891 (1966).
29. W. Gilbert and B. Müller-Hill, *Proc. Natl. Acad. Sci.* **58,** 2415 (1967).
30. J. R. Cann and W. B. Goad, *J. Biol. Chem.* **240,** 148 (1965).
31. L. G. Longsworth and C. F. Jacobsen, *J. Phys. Colloid Chem.* **53,** 126 (1949).
32. R. A. Phelps and J. R. Cann, *J. Am. Chem. Soc.* **78,** 3539 (1956).
33. J. R. Cann and R. A. Phelps, *J. Am. Chem. Soc.* **79,** 4672 (1957).
34. J. R. Cann, *J. Am. Chem. Soc.* **80,** 4263 (1958).
35. J. R. Cann, *J. Phys. Chem.* **63,** 210 (1959).
36. J. R. Cann and R. A. Phelps, *J. Am. Chem. Soc.* **81,** 4378 (1959).
37. J. R. Cann, *J. Biol. Chem.* **235,** 2810 (1960).
38. J. R. Cann, *J. Am. Chem. Soc.* **83,** 4784 (1961).
39. J. R. Cann and W. B. Goad, *Arch. Biochem. Biophys.* **108,** 171 (1964).
40. J. R. Cann and W. B. Goad, *J. Biol. Chem.* **240,** 1162 (1965).
41. J. R. Cann, *Biochemistry* **5,** 1108 (1966).
42. J. R. Cann and W. B. Goad, *Advan. Enzymol.* **30,** 139 (1968).
43. W. C. Parker and A. G. Bearn, *Nature* **199,** 1184 (1963).
44. R. J. Kossman, D. C. Fainer, and S. H. Boyer, *Symp. Quant. Biol. Cold Spring Harbor* **29,** 375 (1964).
45. W. B. Goad and J. R. Cann, *Ann. N.Y. Acad. Sci.* **164,** 172 (1969).
46. J. Hearst and J. Vinograd, *Arch. Biochem. Biophys.* **92,** 206 (1961).
47. J. Rosenbloom and V. N. Schumaker, *Biochemistry* **2,** 1206 (1963).
48. J. B. T. Aten and J. A. Cohen, *J. Mol. Biol.* **12,** 537 (1965).
49. H. Fujita, "Mathematical Theory of Sedimentation Analysis," Chapter IV. Academic Press, New York, 1962.
50. K. Kakiuchi, S. Kato, A. Imanishi, and T. Isemura, *J. Biochem.* **55,** 102 (1964).
51. J. A. Gordon and M. Ottesen, *Biochim. Biophys. Acta* **75,** 453 (1963).
52. J. R. Cann and R. A. Phelps, *J. Am. Chem. Soc.* **77,** 4266 (1955).
53. R. W. Briehl, *J. Biol. Chem.* **238,** 2361 (1963).
54. M. H. Klapper, G. H. Barlow, and I. M. Klotz, *Biochem. Biophys. Res. Commun.* **25,** 216 (1966).
55. M. H. Klapper and I. M. Klotz, *Biochemistry* **7,** 223 (1968).

V/NUMERICAL METHODS[1]

The high-speed electronic computer makes it feasible to simulate transport processes and interactions to a high degree of accuracy; the problem is to obtain sufficient accuracy with a suitably limited quantity of computation. In this chapter, the considerations will be set forth that lead to our formulation of the numerical computation, and in the appendix, a code for sedimentation calculations is presented in FORTRAN language. For an approach rather different from ours, see Dishon et al. (1).

At the outset, we suppose the transport cell to be divided into a number of discrete segments. The basic variables will be the average concentrations in each segment of the various materials being transported. We must accurately relate the currents between segments due to transport processes to these variables. Then, given the distribution of material at one instant, we can compute its change during the next short interval of time; then, from the changed distribution of material, the change over the next short interval, and so on, constructing the entire evolution of the distribution of material in the cell from its initial condition.

We will deal only with the limiting case of a reaction so fast that, in effect, there is local equilibrium among the interacting species. To be definite, the reaction will be taken to be

$$m\mathrm{M} + n\mathrm{X} \rightleftharpoons (\mathrm{M}_m \mathrm{X}_n) \tag{268}$$

A macromolecule M aggregates reversibly into an m-mer M_m with the mediation of a small molecule X, of which n are bound into the complex $\mathrm{M}_m \mathrm{X}_n$.

---

[1] Chapter V was written by Walter B. Goad.

The following notation will be used in this chapter:

r	position in the transport cell
t	time
$C_1(r, t)$	concentration of M, the macromonomer
$C_2(r, t)$	concentration of the complex $M_m X_n$
$C_3(r, t)$	concentration of the small ligand X
$C_L(r, t)$	total concentration of macromolecule $= C_1(r, t) + mC_2(r, t)$
$C_S(r, t)$	total concentration of small ligand $= C_3(r, t) + nC_2(r, t)$
$V_1(r), V_2(r), V_3(r)$	velocities at which the three species are driven by the applied field
D_1, D_2, D_3	diffusion coefficients of the three species
$S(r)$	cross-sectional area of the transport cell at r
r_i	positions of the interfaces between discrete segments into which the transport cell is divided for computational purposes ($i = 1, 2, 3, \ldots, N$)
$\bar{C}_k(i, t)$	average concentration of species k in the segment bounded by r_{i-1}, r_i ($k = 1, 2, 3, L, S$)
$\tau(i)$	volume of the segment r_{i-1}, r_i

The transport parameters $V(r)$ and D may depend on the concentrations, but we will not make this explicit. In a centrifuge cell, V is a sedimentation coefficient multiplied by $\omega^2 r$; in an electrophoresis cell, it is a mobility multiplied by an electric field strength. In a sector-shaped centrifuge cell, $S(r)$ is proportional to r; it is constant in most electrophoresis cells.

From the definitions above,

$$C_L(r, t) = C_1(r, t) + mC_2(r, t), \qquad C_S(r, t) = C_3(r, t) + nC_2(r, t) \tag{269}$$

$$\bar{C}_k(r, t) = \tau(i)^{-1} \int_{r_{i-1}}^{r_i} S(r') C_k(r', t)\, dr' \tag{270}$$

We will take the quantities $C_k(r, t)$ to represent the exact solutions of the transport equations:

$$\frac{\partial C_L(r, t)}{\partial t} = -\frac{1}{S(r)} \frac{\partial}{\partial r} S(r) \left[V_1(r) C_1(r, t) + m V_2(r) C_2(r, t) - D_1 \frac{\partial C_1(r, t)}{\partial r} - m D_2 \frac{\partial C_2(r, t)}{\partial r} \right] \tag{271a}$$

$$\frac{\partial C_S(r, t)}{\partial t} = -\frac{1}{S(r)} \frac{\partial}{\partial r} S(r) \left[V_3(r) C_3(r, t) + n V_2(r) C_2(r, t) - D_3 \frac{\partial C_3(r, t)}{\partial r} - n D_2 \frac{\partial C_2(r, t)}{\partial r} \right] \tag{271b}$$

Numerical Methods

The (fast) reaction (268) requires that, given $C_L(r, t)$ and $C_S(r, t)$, $C_1(r, t)$, $C_2(r, t)$, $C_3(r, t)$ be determined by the system of equations comprising the two equations (269) plus an equilibrium relation:

$$C_1(r, t)^m C_3(r, t)^n = K C_2(r, t) \tag{272}$$

(in which we have assumed complete cooperativity in the aggregation process).

To obtain equations for the average concentrations, we multiply equations (271a, b) by $S(r)$, integrate over a discrete interval (r_{i-1}, r_i), and divide by $\tau(i)$:

$$\frac{\partial \bar{C}_L(i, t)}{\partial t} = \tau(i)^{-1} \{ S(r_{i-1}) [J_V^L(i-1, t) + J_D^L(i-1, t)]$$

$$- S(r_i)[J_V^L(i, t) + J_D^L(i, t)] \} \tag{273a}$$

$$\frac{\partial \bar{C}_S(i, t)}{\partial t} = \tau(i)^{-1} \{ S(r_{i-1}) [J_V^S(i-1, t) + J_D^S(i-1, t)]$$

$$- S(r_i)[J_V^S(i, t) + J_D^S(i, t)] \} \tag{273b}$$

where

$$J_V^L(i, t) = V_1(r_i) C_1(r_i, t) + m V_2(r_i, t) C_2(r_i, t) \tag{274}$$

is the current (total monomer units per unit area per unit time) of macromolecule driven between segments i and $i+1$; positive if from i to $i+1$, negative if from $i+1$ to i;

$$J_V^S(i, t) = V_3(r_i) C_3(r_i, t) + n V_2(r_i, t) C_2(r_i, t) \tag{275}$$

is the corresponding current of small ligand;

$$J_D^L(i, t) = -D_1 \left.\frac{\partial C_1(r, t)}{\partial r}\right|_{r=r_i} - m D_2 \left.\frac{\partial C_2(r, t)}{\partial r}\right|_{r=r_i} \tag{276}$$

is the macromolecular current due to diffusion; and

$$J_D^S(i, t) = -D_3 \left.\frac{\partial C_3(r, t)}{\partial r}\right|_{r=r_i} - n D_2 \left.\frac{\partial C_2(r, t)}{\partial r}\right|_{r=r_i} \tag{277}$$

is the diffusion current of small ligand.

The computation will be in terms of the variables $\bar{C}_k(i, t)$, so we must relate the concentrations and their derivatives at the interfaces r_i to adjacent averages. This begins by making the approximation that $\bar{C}_1(i, t)$, $\bar{C}_2(i, t)$, and $\bar{C}_3(i, t)$ are the values obtained by enforcing equilibrium with $\bar{C}_L(i, t)$ and $\bar{C}_S(i, t)$ given; accuracy then requires that the cell be divided at least finely enough that, in the exact solution, the equilibrium does not shift very much over any segment.

For the next few paragraphs, the subscripts 1, 2, and 3 will be dropped, the argument applying equally to all three.

The relation between concentration at a particular point r and values nearby can be expressed by a Taylor expansion about r_i:

$$C(r, t) = C(r_i, t) + \frac{\partial C}{\partial r}\bigg|_{r_i, t} (r - r_i) + \frac{\partial^2 C}{\partial r^2}\bigg|_{r_i, t} \frac{(r - r_i)^2}{2!}$$
$$+ \frac{\partial^3 C}{\partial r^3}\bigg|_{r_i, t} \frac{(r - r_i)^3}{3!} + \cdots$$

which, by multiplying by $S(r)$, integrating from r_{j-1} to r_j, and dividing by τ_j, leads to a relation between $\bar{C}(j, t)$ and the concentration and its derivatives at r_i:

$$\bar{C}(j, t) = C(r_i, t) + \frac{\partial C}{\partial r}\bigg|_{r_i, t} \langle\delta(r - r_i)\rangle_{\text{av}}^j + \frac{\partial^2 C}{\partial r^2}\bigg|_{r_i, t} \frac{\langle\delta^2(r - r_i)\rangle_{\text{av}}^j}{2!} + \cdots \tag{278}$$

where

$$\langle\delta^n(r - r_i)\rangle_{\text{av}}^j = \tau^{-1} \int_{r_{j-1}}^{r_j} S(r)(r - r_i)^n r$$

are moments of the average displacement of segment j from interface i.

To compute $J_V(i, t)$, we adopt an approximate expression of the form

$$C(r_i, t) = x_1 \bar{C}(i, t) + x_2 \bar{C}(i + 1, t) + x_3 \bar{C}(i - 1, t) + \cdots \tag{279}$$

and determine the x's by using as many terms of the expansion (278) as possible.

The concentration at r_i, $C(r_i, t)$, will be most closely related to the average concentration in the segment just upstream, from which the material crossing interface i is coming. It is essentially for this reason, as will be discussed later, that numerical stability of the equations requires that in Eq. (279) the weight of terms representing segments upstream of r_i dominate.

Accordingly, if only one term is used, we must have $x_1 = 1$ and $x_2 = x_3 = 0$ if $V(r_i) > 0$, and $x_2 = 1$ and $x_1 = x_3 = 0$ if $V(r_i) < 0$. The expansion (278) tells us that, in this case, an error $(\partial C/\partial r)\,\delta r$ is made at each interface; this being of the same form as the diffusion current, the solutions behave as if an additional diffusive process were present with diffusion coefficient proportional to $V\,\delta r$. It is economical of computation to use more terms in the expression for J_V rather than make δr very small. We choose a three-term expression:

$$C(r_i, t) = x_1^F \bar{C}(i, t) + x_2^F \bar{C}(i + 1, t) + x_3^F \bar{C}(i - 1, t) \quad \text{if} \quad V(r_i) > 0$$
$$C(r_i, t) = x_1^B \bar{C}(i + 1, t) + x_2^B \bar{C}(i, t) + x_3^B \bar{C}(i + 2, t) \quad \text{if} \quad V(r_i) < 0$$
$$\tag{279'}$$

Numerical Methods

Substituting the expansions of $\bar{C}(i-1, t)$, $\bar{C}(i, t)$, $\bar{C}(i+1, t)$, Eq. (278), into Eq. (279′) for $V(r_i) > 0$, we obtain

$$C(r_i, t) = (x_1^F + x_2^F + x_3^F)\, C(r_i, t)$$

$$+ \frac{1}{2} \frac{\partial C}{\partial r}\bigg|_{r_i, t} [-x_1^F(r_i - r_{i-1}) + x_2^F(r_{i+1} - r_i)$$

$$+ x_3^F(r_{i-1} + r_{i-2} - 2r_i)]$$

$$+ \frac{1}{6} \frac{\partial^2 C}{\partial r^2}\bigg|_{r_i, t} \{x_1^F(r_i - r_{i-1})^2 + x_2^F(r_{i+1} - r_i)^2 + x_3^F[(r_{i-1} - r_i)^2$$

$$+ (r_{i-1} - r_i)(r_{i-2} - r_i) + (r_{i-2} - r_i)^2]\} + O(\delta r^3)]$$
(280)

Then, if x_1^F, x_2^F, x_3^F are made to satisfy the two equations obtained by making the coefficients of $\partial C/\partial r$ and $\partial^2 C/\partial r^2$ equal to zero, plus the equation

$$x_1^F + x_2^F + x_3^F = 1, \tag{281}$$

J_V will be accurate to order δr^3. The result is

$$x_3^F = (r_i - r_{i-1})(r_{i+1} - r_i)/(r_{i-2} - r_i)(r_{i+1} - r_{i-2})$$
$$x_2^F = (r_i - r_{i-1})(r_i - r_{i-2})/(r_{i+1} - r_{i-1})(r_{i+1} - r_{i-2}) \tag{282}$$
$$x_1^F = 1 - x_2^F - x_3^F$$

and there are corresponding expressions for the case $V(r_i) < 0$.

For the diffusion currents, we need $\partial C/\partial r$ at the interfaces. We set

$$\frac{\partial C}{\partial r}\bigg|_{r_i} = y_1 \bar{C}(i, t) + y_2 \bar{C}(i+1, t) \tag{283}$$

and apply the analogous argument, with the result

$$y_2 = -y_1 = 2/(r_{i+1} - r_{i-1}) \tag{283′}$$

and $J_D(i, t)$ is accurate to order δr^2.

At the boundaries, $i = 1$ and $i = N$, we shall usually want both diffusion and driven currents to be zero. The three-term expression for J_V must also be modified at $i = 2$ if $V(2, t) > 0$ and at $i = N - 1$ if $V(N - 1, t) < 0$, for it would otherwise involve segments outside the cell. In either case, we make $x_3 = 0$; by making the expansion (278) satisfy the Archibald condition at the boundary, $D(\partial C/\partial r) = VC$ (that is, by assuming that sufficiently near boundaries, the rate of change of the concentration can be neglected as compared with either of the nearly balanced diffusion and driven currents), one has only two equations for x_1 and x_2 to satisfy to make J_V accurate to order δr^3. Here, the x's are solutions of the equations

$$x_1^F[1 - (r_i - r_{i-1})^2 A] + x_2^F[1 - (r_{i+1} - r_i)^2 A] = 1$$

$$x_1^F[r_{i-1} - r_i + (r_i - r_{i-1})^2 B] + x_2^F[r_{i+1} - r_i + (r_{i+1} - r_i)^2 B] = 0$$

where

$$A = (V/D)\{6(r_i - r_{i-1})[1 + (V/D)\tfrac{1}{2}(r_i - r_{i-1})]\}^{-1}$$

$$B = [(V/D)(r_i - r_{i-1}) + 1]\{3(r_i - r_{i-1})[1 + (V/D)\tfrac{1}{2}(r_i - r_{i-1})]\}^{-1}$$

and $x_3^F = 0$. Again, a similar expression can be obtained for $V < 0$.

When the approximations for the concentrations and their gradients are substituted into the expressions (274)–(277) for the currents, and these, in turn, are introduced into the transport equations (273), we have formulas for the time rate of change of the average concentrations $\bar{C}(i, t)$ at each i in terms of linear combinations of the values in neighbouring segments. So, given values at any t, we can calculate their rate of change, and, by extrapolating over a short interval Δt, obtain the values at $t + \Delta t$, recalculate the equilibrium, recalculate the currents, and from the time rates of change, extrapolate over the next Δt, and so on.

This is the simplest way of proceeding. It corresponds to stopping after the second term in a Taylor expansion

$$\bar{C}(t + \Delta t) = \bar{C}(t) + (\partial \bar{C}/\partial t)\,\Delta t + \tfrac{1}{2}(\partial^2 \bar{C}/\partial t^2)\,\Delta t^2 + \cdots$$

Thus, we introduce an error of order Δt^2 at each step for $t/\Delta t$ steps to calculate the evolution up to t. The overall errors, of order Δt from the time differencing and order δr^3 and δr^2 from our approximation to J_V and J_D, respectively, are usually called truncation errors, and can be made appropriately small by choosing Δt and δr sufficiently small. A more complicated issue is that of numerical stability, which requires a relation between Δt and δr, no matter how small they may be.

The essence of stability in the problem at hand is this: our equations describe a continuous physical process in which the migration of material during each infinitesmal interval of time produces changes that determine its migration during the next infinitesmal interval. For computational purposes, we replace this process with a discrete one. In a region where the concentration is roughly $\bar{C}$, during Δt the overall motion of material is made up of discrete transfers at each of a number of segment interfaces in which amounts of material proportional to $\bar{C}V\,\Delta t$ are moved by the applied field, and amounts roughly proportional to $\bar{C}D\,\Delta t/\delta r$ are moved by diffusion. On the same scale, the discrete amounts of material whose concentrations will be influenced by the motion are proportional to $\bar{C}\delta r$. If the elements of change in the discrete process become comparable with the elements being changed, we have a process rather different than the continuous one we set out to simulate. For given δr, as Δt is increased, somewhere in the vicinity where $V \Delta t + D \Delta t/\delta r = \delta r$ a transition in behavior occurs, beyond which the solutions for the discrete process depart ever more rapidly as time goes on from those for the continuous process; in fact, the difference grows exponentially in t. In this, the unstable regime, the discrete and continuous processes represent different phenomena. Instability is avoided by choosing Δt sufficiently small to be comfortably within the stable regime.

These remarks apply only if the currents have been computed so as to allow the concentrations to determine the transport in a physical way, and this is behind the earlier remark that, in constructing J_V, the upstream values of the average concentration must dominate. If not, one has a discrete process that is unstable for every value of Δt, no matter how small.

For more technical discussions of stability, and of the results and limitations of existing analyses, the reader is referred to the book by Richtmyer and Morton (2).

The continuous process could be more accurately represented. For example, the currents could be taken to vary linearly with time between their values at the beginning and end of each interval; then our equations would have the form

$$\bar{C}(i, t + \Delta t) = \bar{C}(i, t) + \tau(i)^{-1}\{S(r_{i-1},)\tfrac{1}{2}[J(i-1, t+\Delta t) + J(i-1, t)]$$
$$- S(r_i)\tfrac{1}{2}[J(i, t+\Delta t) + J(i, t)]\}$$

(we have let J represent the total current, $J_V + J_D$). Now, since, through the J's, the equations involve values of $\bar{C}(i', t + \Delta t)$ at segments i' that flank segment i, the equation at each i contains as unknowns the average concentrations at several i', and a system of linear equations must be solved, instead of just computing $\bar{C}(i, t + \Delta t)$ in terms of the $\bar{C}(i', t)$ as before.

These equations are said to be *implicit*, and, apart from possible effects of reequilibrating the species in discrete steps, they are unconditionally stable. In addition, the truncation error due to time differencing is reduced to order Δt^3 per step. This is paid for in computational complexity, solving an $N \times N$ system of linear equations at each step. The system is, however, very well conditioned, and the solution is readily accomplished by Gaussian elimination.

We must have $V \Delta t/\delta r + D \Delta t/\delta r^2 < 1$ if the explicit equations are used, and for sufficiently small δr, Δt is determined by the diffusion term. Since it seems very complicated to make the equations completely implicit, including the (nonlinear) reequilibration, we must, in any case, make Δt sufficiently small that the equilibrium is not much shifted in any segment by transport during a time interval. Our experience is that in these circumstances it is advantageous to make the diffusion terms implicit, but still meet a rather tight criterion, $V \Delta t/\delta r < \frac{1}{4}$, so that there is little advantage in making the terms in J_V implicit. The resulting system can be solved by the method described by Richtmyer and Morton (3) for the heat-flow problem; it is a very convenient codification of Gaussian elimination.

The truncation errors that depend on δr are especially severe where the concentration varies rapidly along the transport cell—where, unfortunately, we want the most accuracy of detail. This difficulty is ameliorated by introducing a moving coordinate system: Then, the segment spacing can be made fine enough for detail where it is needed and coarser elsewhere, with the closely spaced region made to move down the transport cell with the front of rapidly varying concentration. A further advantage is that the transport coefficient $V(r_i)$ is replaced by $V(r_i) - V_c(i)$, where $V_c(i)$ is the velocity of the coordinate system at i, further reducing the truncation error in J_V. The cost of this is added complexity of the code for transport cells whose ends must be taken account of.

For the centrifuge cell, we move each segment interface, excepting those at the ends, with the velocity of the less rapidly sedimenting macromolecular species. When the coordinate spacing at the top, or meniscus, of the cell exceeds an assigned value, a new segment is introduced, initially very small and empty of solute. It grows until its assigned size is reached (which, if the spacing is variable, will differ from that of its predecessor), when another small empty segment is introduced, and so on. Meanwhile, at the bottom of the cell, segments are being removed from the calculation: The next-to-last interface, which is the bottommost moving one, is made to sweep up material in the last segment and transfer it to the next-to-last one. This is accomplished by adding a term $V_c(N-1)\bar{C}(N, t)S(r_{N-1})$ to J_V and J_D to form the total current at that interface. When r_{N-1} reaches the bottom, all material formerly in the bottom segment has been transferred to the next-to-bottom one and the empty segment is dropped from the calculation by renumbering all of the segments.

To summarize, during each Δt, the radii are advanced at a prescribed velocity $V_c(i)$ and the changes computed in the average concentrations $\bar{C}_1(i, t)$, $\bar{C}_2(i, t)$, and $\bar{C}_3(i, t)$ due to changes in the segment volumes $\tau(i)$ associated with movement of the coordinate system, and due to driven transport and diffusion during the interval. The principal approximations are: (1) there is no reequilibration by interaction during Δt—each component migrates independently from its distribution at the beginning of the interval; (2) the current due to driven migration of each component is constant during Δt and is determined by its distribution at the beginning of the interval; (3) migration by diffusion takes place as if the change in the distribution of each component were linear in time during Δt; and (4) each term in the transfer between segments suffers from errors due to truncation of the series expansion, Eq. (278).

At each time step, after the concentrations have been advanced, equilibrium is recalculated. That is, for known $\bar{C}_L(i, t)$ and $\bar{C}_S(i, t)$ computed from $\bar{C}_1(i, t)$, $\bar{C}_2(i, t)$, and $\bar{C}_3(i, t)$ as changed by transport during Δt, it is necessary to find new values of $\bar{C}_1$, $\bar{C}_2$, and $\bar{C}_3$ that satisfy Eq. (269) and (272) simultaneously.

We eliminate $\bar{C}_1$ and $\bar{C}_3$ and obtain an equation for $w = m\bar{C}_2/\bar{C}_L$, the proportion of macromolecular material complexed:

$$w - (mK/\bar{C}_L)[\bar{C}_L(1-w)]^m \{\bar{C}_S[1 - (n\bar{C}_L/m\bar{C}_S)w]\}^n = 0 \qquad (284)$$

We solve it by an iterative procedure.

Straightforward application of Newton's method is impractical if either n or m is large, for it then converges very slowly. At $n = 100$, for example, many thousands of iterations may be required to advance the accuracy of the solution from 1 to 0.1%. Instead, we do the following: Let the left-hand side of Eq. (284) be $Y(w)$, whose zero we seek. Schematically, it will look something like the function plotted in Fig. 90. Newton's method consists of starting with an estimate of w, w^0, computing the tangent to the curve at w^0, taking the zero of the tangent as an improved estimate, w^1, repeating the process to obtain w^2, and so on, until $Y(w^n)$ is sufficiently close to zero. Slow convergence is due to extreme curvature in $Y(w)$.

Referring now to Fig. 90, our method is based on the following observations. Any two values of w, w_{min}^0 and w_{max}^0, such that $Y(w_{min}^0) < 0$ and $Y(w_{max}^0) > 0$, are limits between which the solution must lie; the zero of the secant AB furnishes an improved lower limit w_{min}^1, and the zero of one of the tangents, either BD or AC, provides an improved upper limit w_{max}^1. With determination of each new pair (w_{min}^n, w_{max}^n) the construction is repeated, until the limits lie sufficiently close to each other.

The number of iterations required can be minimized by initiating the procedure with a value of w estimated from a solution previously obtained. At each i, the solution $w_{old}(i)$ that corresponds to $\bar{C}_L(i, t)$ and $\bar{C}_S(i, t)$ is saved in memory. The value of w for starting the equilibration calculation at $(i, t + \Delta t)$ is computed as

$$w^0 = w_{old}(i) + \frac{\partial w}{\partial C_L}\bigg|_{w_{old}(i)} [\bar{C}_L(i, t + \Delta t) - \bar{C}_L(i, t)]$$

$$+ \frac{\partial w}{\partial C_S}\bigg|_{w_{old}(i)} [\bar{C}_S(i, t + \Delta t) - \bar{C}_S(i, t)]$$

The partial derivatives are obtained by differentiating Eq. (284) with respect to $\bar{C}_L$ and $\bar{C}_S$ and solving for them as functions of w, $\bar{C}_L$, and $\bar{C}_S$. Their values

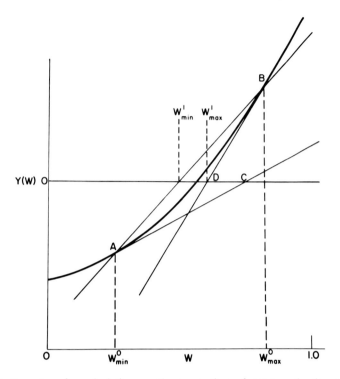

FIG. 90. Procedure for calculating w, the proportion of macromolecular material complexed [see Eq. (284)].

are computed as each solution is obtained and saved in memory for use in the equation above at the next time step.

The reaction is also of interest in which small ligand is bound to the macromonomer, but must be freed upon formation of the macromolecular complex:

$$m(MX_n) \rightleftharpoons M_m + mnX \tag{285}$$

With complete cooperativity, the equilibrium relation is

$$C_1(r, t)^m = KC_2(r, t)C_3(r, t)^{mn} \tag{286}$$

This relation can be put into the same form as Eq. (272) by taking the mth root of both sides:

$$C_1(r, t) = K^{1/m} C_2(r, t)^{1/m} C_3(r, t)^n$$

which is to be taken simultaneously with the equations

$$C_L(r, t) = C_1(r, t) + mC_2(r, t) \quad \text{and} \quad C_S(r, t) = C_3(r, t) + nC_1(r, t)$$

and the method just described is applied without change.

A number of examples of numerical solutions computed by the methods described in this chapter have been presented in Chapter IV.

REFERENCES

1. M. Dishon, G. H. Weiss, and D. A. Yphantis, *Biopolymers* 4, 449 (1966).
2. R. D. Richtmyer and K. W. Morton, "Difference Methods for Initial-Value Problems," 2nd ed. Wiley (Interscience), New York, 1967.
3. R. D. Richtmyer and K. W. Morton, "Difference Methods for Initial-Value Problems," 2nd ed., pp. 198–201. Wiley (Interscience), New York, 1967.

APPENDIX: A COMPUTER CODE FOR SEDIMENTATION CALCULATIONS

The code is presented in FORTRAN language as accepted by compilers of Control Data Corporation's CHIPPEWA operating system for its 6000 series computers.

In the following notes, variables will be denoted by their symbolic FORTRAN names as in the code.

1. Input data consists of an identifying remark (1 card) plus specification of coordinate system, specification of the reaction, initial concentrations, transport coefficients, speed of rotation in rpm, specification of number of print cycles, and the point to which the problem is to be run. The last is specified as the fraction of the distance down the centrifuge cell that the macromonomer is to sediment before the problem is terminated.

2. If NUMSMA is positive, the reaction is that of Eq. (268), with $n =$ NUMSMA, $m =$ MMER; if negative, the reaction is that of Eq. (285), with $n = -$NUMSMA, $m =$ MMER. In the first case, the subroutine EQUIL is used; in the second, the subroutine EQUIL2. The latter differs from the former only in a number of exchanges of the symbolic names of quantities in storage.

3. Two coordinate systems are available: If ISPACE = 1, uniformly logarithmic spacing is provided; if ISPACE > 1, there is a closely spaced region with minimum spacing initially at the meniscus. The spacing increases according to a Gaussian distribution with half-width specified by the value of the parameter ISPACE. The ratio of maximum to minimum spacing is specified by the value of the fourth parameter on the second card, T.

4. Initial concentrations are specified and the required equilibrium constant computed from them.

```
      PROGRAM SEDMNT(INPUT,OUTPUT)
      COMMON C1( 101),C2( 101),C3( 101),CL( 101),CS( 101),R( 101),
     1DXL(101),DXS(101)
      COMMON/TEMP/DEL( 101),DELMS( 101),V ( 101),VS( 101),FT1(101),FT1(1
     101)
      COMMON/PARAM/ISPACE,MMER,NUMSMA,KM,KN,D1,D2,D3,ESTAB,FLM,FLN,C1I,
     1C2I,C3I,S1I,S2I,S3I,RPM,    S1,S2,S3,DEL1,DELR,RMEN,RBOT,TIME,IMIN,
     2DELT,K1,D1T,D2T,FDN,EQUEPS,D3T ,DELS,WMAX,WMIN
      COMMON/FODATA/WLAST(101),CM(101),CN(101)
      COMMON/COORD/KRMEN,X1F(101),X2F(101),X3F(101),X1B(101),X2B(101),X3
     1B(101),DIP(101),DIM(101),DTEMP
      COMMON/EQPAR/CLT,CST,IQ,IT,REM,FNM
      REAL KN,KM,LARMOL,NEXTPP,K1
      DATA ESTAB/.25/,  EQUEPS/.001/,    ENDDEL/.10/,RMEN/5.9/,
     1RBOT/7.4/
      COMMON/FLAG/IZ
      DIMENSION RMRK(7)
    5 READ 106,RMRK
  106 FORMAT(7A10)
      READ  104,ISPACE,NUMSMA,MMER,T
      READ  108,                      C1I,C2I,C3I,S1,S2,S3,D1,D2,D3,RPM,
     1PRINT1,FINISH
  104 FORMAT(3I12,F12.0)
  108 FORMAT(6F12.0/6F12.0)
      NSGN= NUMSMA
      IF(ISPACE.EQ.0) GO TO 8
      IZ=1 $ IF(NUMSMA.GT.0)  GO TO 16 $ IZ=2 $ NUMSMA=-NUMSMA
   16 ILOG=ISPACE $ ISPACE=100
      FLN=FLOAT(NUMSMA)
      FLM=FLOAT(MMER)
      FNM=FLN/FLM $ REM=1.0/FLM
      FDN=0
      IT=0
      IF(ILOG.GT.1) GO TO 4
      IMIN=4          $ ITOT=ISPACE-IMIN+1 $ FPZ=T=DSQ=0
      A=(RBOT/RMEN)**(1.0/FLOAT(ITOT))    $ GO TO 9
    4 TS=0 $ DSQ=1./ILOG**2 $ A=1. $ ILOG=2 $ T=(1.-1./T)*EXP(DSQ)
      DO 13 I=1,48
   13 TS=TS+EXP(-DSQ*I**2) $ TS1=ALOG(RBOT/RMEN) $EPZ=.5*TS1/(48.-TS*T)
      TS=.5*TS1 $ DO10 I=49,100 $ TS=TS+FPZ*(1.-EXP(-DSQ*I**2)*T)
   10 IF(TS.GE.TS1) GO TO 14
   14 IMIN=ISPACE-I+1
    9 R(IMIN-1)=RMEN $ DO 1 I=IMIN,ISPACE
    1 R(I)=R(I-1)*A*EXP( FPZ*(1.-EXP(-DSQ*(I-IMIN+1)**2)*T))
      DEL1=R(IMIN)-RMEN $ R(ISPACE)=RBOT  $ KRMEN=0
      GO TO (21,22),IZ
   22 K1=C1I $ KM=1.0/C2I $ SMAMOL=C3I+FLN*C1I/FLM $ GO TO 23
   21 K1=C2I
      KM=1.0/C1I
      SMAMOL=C3I+FLN*C2I
   23 KN=1.0/C3I
      LARMOL=C1I+FLM*C2I
      PRINT 107,RMRK
  107 FORMAT(1H17A10)
      PRINT 105,LARMOL,SMAMOL,MMER,NSGN,          C1I,C2I,C3I,S1,S2,S3,
     1D1,D2,D3,RPM
  105 FORMAT(14H0 INITIAL DATA/35H0TOTAL MACROMOLECULE CONCENTRATION=F10
     1.2,35H0TOTAL SMALLMOLECULE CONCENTRATION=F10.2,3X12HREACTION M2=I3
     2,4H(M1)I5,4H(M3)/23H0INITIAL CONCENTRATIONS6X3E15.4/29H0SEDIMENTAT
     3ION COEFFICIENTS  3E15.4/23H0DIFFUSION COEFFICIENTS6X3F15.4/5H0RPM
     4=F10.2)
      S1I=S1
      S2I=S2
      S3I=S3
      DST=((RPM*1.0E-4)**2)*1.0966195E-7
      S1=S1*DST
      S2=S2*DST
      S3=S3*DST
      DST=AMIN1(S1,S2)
```

Appendix: A Computer Code for Sedimentation Calculations 219

```
      FINISH=ALOG(1.0+FINISH*(RROT/RMEN-1.0))/DST
      PRINTI=FINISH/PRINTI
            D1T=D1*2. $ D2T=D2*2. $ D3T=D3*2.
      DFLS=1. $ DELT=.4*DEL1 /(S2*RMEN)
      TIME=0
      NEXTPR=PRINTI
      GO TO (25,26),IZ
   25 B=FLN*LARMOL/(FLM*SMAMOL)
      W=FLM*C2T/LARMOL
      DT=1.0+W*(FLM/(1.- W)+B*FLN/(1.0-B*W)) $ DS=FLN/(DT*SMAMOL)
      DT=(FLM-1.0)/(DT*LARMOL)$ GO TO 27
   26 DS=0 $ DT=0 $ W=C1I/LARMOL $ DST=1.0/FLM
   27 DS=W*DS $ DT=W*DT
      DO 11 I=1,ISPACE $ DXS(I)=DS $ DXL(I)=DT
      WLAST(I)=WSCM(I)=DST*LARMOL$CN(I)=SMAMOL
      CL(I)=LARMOL
      CS(I)=SMAMOL
      C1(I)=C1I
      C2(I)=C2I
   11 C3(I)=C3I
      IMAX=ISPACE-2$ DO 28 I=IMIN,IMAX
      DR=R(I)-R(I-1) $ DP=R(I+1)-R(I) $ DPP=R(I+2)-R(I+1)
      TS=R(I+2)-R(I-1) $ X3F(I+1)=-DP*DPP/((DP+DR)*TS)
      X2F(I+1)=DP*(DP+DR)/((DP+DPP)*TS) $X1F(I+1)=1.-X2F(I+1)-X3F(I+1)
      X3B(I)=-DP*DR/((DP+DPP)*TS) $X2B(I)=DP*(DP+DPP)/((DP+DR)*TS)
      X1B(I)=1.-X2B(I)-X3B(I)
      DIP(I+1)=R(I+1)/(DPP+DP) $ DIM(I+1)=R(I)/(DR+DP)
   28 CONTINUE
      DIP(IMIN)=DIM(IMIN+1)                    $DIM(ISPACE)=DIP(ISPACE-1)
      DIP(ISPACE)=0 $ GO TO 15
    2 DELT=DELT*FSTAB/DFLS $DFLS=0
      DST=DELT*RMEN*S2/DEL1  $ IF(DST.GT..5) DELT=.5*DELT/DST
      DST=RROT*S2*DELT/(R(ISPACE-1)-R(ISPACE-2))
      IF(DST.GT..10) DELT=DELT*.10/DST
      CALL RADIUS
      IF(ILOG .EQ.1) GO TO 3 $ IF(KRMEN.EQ.0) GO TO 3
      DEL1=RMEN*(EXP(FPZ*(1.-EXP(-DSQ*ILOG**2)*T))-1.)
      ILOG=ILOG+1 $ KRMEN=0
    3 IT=IT+1
      TIME=TIME+DELT
      IF(NEXTPR.GT.TIME) GO TO 2
      NEXTPR=TIME+PRINTI
   15 TOT1=0 $ TOT2=0 $ TOT3=0 $ TOT4=0 $ TOT5=0
      DO12 I=IMIN,ISPACE
      DST=(R(I)-R(I-1))*(R(I)+R(I-1))
      TOT4=TOT4+CL(I)*DST
      TOT5=TOT5+CS(I)*DST
      TOT1=TOT1+C1(I)*DST
      TOT2=TOT2+C2(I)*DST
   12 TOT3=TOT3+C3(I)*DST
      TOTL=TOT1+FLM*TOT2
      TOT5=TOT3+FLN*TOT2$ IF(IZ.EQ.2) TOT5=TOT3+FNM*TOT1
      PRINT 100,   TIME,DELT,TOT1,TOT2,TOT3,TOTL,TOT5 ,TOT4,TOT5
      FAN=FDN/FLOAT(IT)
      PRINT 102,    IT,FAN
      RB=RMEN $ ISM=ISPACE-1 $ DO 6 I=IMIN,ISM
      DEL(I)=2.*(CL(I+1)-CL(I))*DIP(I)/R(I)  $ RB=R(I)
      PRINT 103,    I,R(I),DEL(I),    C1(I),C2(I),C3(I),CL(I),CS(I)
    6 CONTINUE
  103 FORMAT(1H ,I4,F12.8,6F12.5)
    7 IF(TIME.LT.FINISH)GOTO2
  102 FORMAT(8HOENTRYS=I6,6H NBAR=F18.4)
      GO TO 5
    8 CALL EMPTY $ STOP
  100 FORMAT(14HOPLOT AT TIME F10.3,15H SECONDS(DELTA=F11.4
     1,     9H SECONDS)//97H TOTAL MOLECULE 1  TOTAL MOLECULE 2  TOTAL MO
     1LECULE 3  TOTAL LARGE MOLECULE  TOTAL SMALL MOLECULE/1H 3E18.6,2E2
     22.6/1H 54X2F22.6//)
      END
```

```
      SUBROUTINE RADIUS
      COMMON/FLAG/IZ
      COMMON/FQDATA/WLAST(101),CM(101),CN(101)
      COMMON/TEMP/FTS(101),FTL(101),FTS(101),FTL(101),FT1(101),FT1(101)
      COMMON C1( 101),C2( 101),C3( 101),CL( 101),CS( 101),R( 101),
     1DXL(101),DXS(101)
      COMMON/FQPAR/CLT,CST,IQ,IT,REM,FNM
      COMMON/PARAM/IS,M,N,KM,KN,DDUM(3) ,ESTAB,FLM,FLN,CDUM(6),
     1SPFED,S1,S2,S3,    DEL1 ,DELR ,RMEN,RBOT,TIME,IMIN,
     2DELT,K1,D1 ,D2 ,FDN,FQUEPS,D3  ,STAB,WMAX,WMIN
      COMMON/COORD/KRMEN,X1F(101),X2F(101),X3F(101),X1B(101),X2B(101),X3
     1B(101),DIP(101),DIM(101),DTEMP
      REAL KM,KN,K1,JPS,JPL,JP1
      FL=DELT*(S2-S1) $ IT=0             $ GL=DELT*S1 $ GS=DELT*S3
      DL=DELT*D2                         $ FL=DELT*D1     $ FS=DELT*D3
      RAD=EXP(GL)
      GMAX=AMAX1(FL,GL-GS)
      IF(R(IMIN).LE.RMEN+DEL1) GO TO 6
      IF(IMIN.EQ.2) GO TO 6
      IMIN=IMIN-1 $ R(IMIN-1) =RMEN $ R(IMIN)=R(IMIN+1)-DEL1
      WLAST(IMIN)=WLAST(IMIN+1)
      C1(IMIN)=C1(IMIN+1) $ C2(IMIN)=C2(IMIN+1) $ C3(IMIN)=C3(IMIN+1)
      CL(IMIN)=CL(IMIN+1)$ CS(IMIN)=CS(IMIN+1)
      DXS(IMIN)=DXS(IMIN+1)$ DXL(IMIN)=DXL(IMIN+1)
      I=IMIN+1
      DR=R(I)-R(I-1) $ DP=R(I+1)-R(I) $ DPP=R(I+2)-R(I+1)
      TS=R(I+2)-R(I-1) $ X3F(I+1)=-DP*DPP/((DP+DR)*TS)
      X2F(I+1)=DP*(DP+DR)/((DP+DPP)*TS) $X1F(I+1)=1.-X2F(I+1)-X3F(I+1)
      DIP(I+1)=R(I+1)/(DPP+DP) $ DIM(I+1)=R(I)/(DR+DP)
      DIM(I+2)=DIP(I+1)
      KRMEN=1
    6 I=IMIN
      DR=R(I)-R(I-1) $ DP=R(I+1)-R(I) $ DPP=R(I+2)-R(I+1)
      DIP(I+1)=R(I+1)/(DPP+DP) $ DIM(I+1)=R(I)/(DR+DP)
      DIP(I)=DIM(I+1) $ DIM(I)=0 $ DIM(I+2)=DIP(I+1)
      G=RMEN*S2/D2 $ TS=G*DR $ TS1=(1.+TS)*3.*DR $ A=G/TS1
      B=2.*(TS+.5)/TS1
      TS=DR*(DR*B-1.) $ TS1=1.-DP*DP*A $ TS2=1.-DR*DR*A
      X2F(I)=TS/(TS1*TS-TS2*DP*(1.+DP*B)) $X3F(I)=0
      X1F(I)=(1.-X2F(I)*TS1)/TS2
      TS=R(I+2)-R(I-1) $ X3F(I+1)=-DP*DPP/((DP+DR)*TS)
      X2F(I+1)=DP*(DP+DR)/((DP+DPP)*TS) $X1F(I+1)=1.-X2F(I+1)-X3F(I+1)
      X3B(I)=-DP*DR/((DP+DPP)*TS) $X2B(I)=DP*(DP+DPP)/((DP+DR)*TS)
      X1B(I)=1.-X2B(I)-X3B(I)
      ISM=IS-1
      DP=R(ISM-1)-R(ISM-2) $ DR=R(ISM)-R(ISM-1)
      G=R(ISM)*S3/D3$TS=G*DR $ TS1=(1.+TS)*3.*DR $ A=G/TS1
      B=2.*(TS+.5)/TS1
      TS=DR*(DR*B-1.) $ TS1=1.-DP*DP*A $ TS2=1.-DR*DR*A
      X2B(ISM-1)=TS/(TS1*TS-TS2*DP*(1.+DP*B)) $ X3B(ISM-1)=0
      X1B(ISM-1)=(1.-X2B(ISM-1)*TS1)/TS2
      CPL=CPS=CML=CMS=CP1=CM1=UML=UMS=0
      K=I-1 $ AMS=0$AML=0 $ FTS(K)=0 $ FTL(K)=0 $ FTL(K)=0 $ ETS(K)=0
      FT1(K)=0 $ FT1(K)=0
      JPL=JPS=AM1=JP1=0
                       RN=R(I)+RMEN $               RM=RMEN
   11 UL= FL *C2(I)     $      US=(GS-GL)*C3(I) $ RN=RM
      RM=R(I+1)-R(I)+RMEN
                       DO 1 I=IMIN,ISM
   13 UPL= FL*C2(I+1)         $ UPS=(GS-GL)*C3(I+1)
      AMS=FS*DIM(I) $ APS=FS*DIP(I) $ AM1= FL*DIM(I) $ AP1=FL*DIP(I)
      AML=DL*DIM(I) $ APL=DL*DIP(I) $ ADS=AMS+APS $ ADL=AML+APL
      AD1=AM1+AP1
      RM=RN $ RN=R(I) $ R(I)=R(I)*RAD
      DR=R(I)-R(I-1)
      IF(I.GT.IMIN)STAB=AMAX1(STAB,GMAX*R(I)/DR)
      IF(I.LT.ISM) GO TO 2 $ IF(R(I).LT.RBOT) GO TO 4
      R(I)=RBOT
    4 TS=(R(I)-RN)*(R(I)+RN)$ JPL=TS*C2(I+1)$JPS=TS*C3(I+1)
```

Appendix: A Computer Code for Sedimentation Calculations

```
      JP1=TS*C1(I+1)
      IF(R(I).EQ.RBOT) GO TO 16
   2                  V=DR*(R(I)+R(I-1))    $ VN=(RN-RM)*(RN+RM)
      D=1.0/(ADS    +V-AMS*FTS(I-1))  $ FTS(I)=APS*D
      IF(I.LT.ISM) GO TO 43 $ CPL=CPS=CP1=0 $ GO TO 44
  43  CPL=X1F(I)*UL+X2F(I)*UPL+X3F(I)*UML
                             IF(I+1.EQ.ISM)CPL=UL $ IF(CPL) 41,41,42
  41  TSO=C2(I+2)*EL $ IF(I+2.EQ.IS) TSO=UL
      CPL=X1B(I)*UPL+X2B(I)*UL+X3B(I)*TSO
  45  IF(CPL.GT.0.) CPL=0
  42  TSO=(GS-GL)*C3(I+2)  $ IF(I+2.EQ.IS) TSO=US
      CPS=TSO*CPS+APS*(C3(I+1)-C3(I))  $ CPL=TSO*CPL+APL*(C2(I+1)-C2(I))
  46  IF(CPS) 48,47,47
  47  CPS=X1F(I)*US+X2F(I)*UPS+X3F(I)*UMS
                             IF(UMS.LE.0.)CPS=US $IF(CPS.LT.0.)CPS=0
  48  TSO=-2.0*R(I)**2 $ CP1=AP1*(C1(I+1)-C1(I))
      CPS=TSO*CPS+APS*(C3(I+1)-C3(I))  $ CPL=TSO*CPL+APL*(C2(I+1)-C2(I))
  44  FTS(I)=D*(C3(I)*VN+JPS+CPS-CMS+AMS*FTS(I-1))
      D=1./(AD1+V-AM1*FT1(I-1)) $ FT1(I)=AP1*D
      FT1(I)=D*(C1(I)*VN+AM1*FT1(I-1)+JP1+CP1-CM1)
      D=1.0/(ADL    +V-AML*FTL(I-1))  $ FTL(I)=APL*D
      FTL(I)=D*(C2(I)*VN+AML*FTL(I-1)+JPL+CPL-CML)
      UML=UL $ UMS=US $ CMS=CPS $ CML=CPL $ UL=UPL   $ CM1=CP1
   1  US=UPS $ GO TO 3
  16  I=IS$ JPS=-JPS+CPS $ JPL=-JPL+CPL $ JP1=-JP1+CP1
   7  R(I)=R(I-1) $ CL(I)=CL(I-1) $ CS(I)=CS(I-1)   $ C1(I)=C1(I-1)
      FTS(I)=FTS(I-1)$FTL(I)=FTL(I-1)$FTS(I)=FTS(I-1)$FTL(I)=FTL(I-1)
      ET1(I)=ET1(I-1) $ FT1(I)=FT1(I-1)    $ WLAST(I)=WLAST(I-1)
      CM(I)=CM(I-1) $ CN(I)=CN(I-1)
      DXS(I)=DXS(I-1) $ DXL(I)=DXL(I-1)
      X1F(I)=X1F(I-1) $ X2F(I)=X2F(I-1) $ X3F(I)=X3F(I-1)
      DIP(I)=DIP(I-1) $ DIM(I)=DIM(I-1)
      X1B(I)=X1B(I-1) $ X2B(I)=X2B(I-1) $ X3B(I)=X3B(I-1)
      C2(I)=C2(I-1) $ C3(I)=C3(I-1)   $ I=I-1 $ IF (I.GE.IMIN) GO TO 7
      IMIN=IMIN+1 $ VN=(RBOT-RM)*(RN+RM) $ GO TO 5
   3  VN=(RBOT-RN)*(RBOT+RN)
   5  V=(RBOT+R(ISM))*(RBOT-R(ISM))
      AMS=DIP(ISM)*ES $ AML=DIP(ISM)*DL $ ADL=DIM(IS)*DL
      ADS=DIM(IS)*FS $ AM1=DIP(ISM)*FL $ AD1=DIM(IS)*FL
      I=IS $ C2T=(VN*C2(I)-JPL+AML*FTL(I-1))/(-AML*FTL(I-1)+ADL+V)
      C3T=(VN*C3(I)-JPS+AMS*FTS(I-1))/(-AMS*FTS(I-1)+ADS+V)
      C1T=(VN*C1(I)-JP1+AM1*FT1(I-1))/(-AM1*FT1(I-1)+AD1+V)
      IF(IZ.EQ.2) GO TO 35 $ CST=C3T+FLN*C2T $ GO TO 36
  35  CST=C3T+FNM*C1T
  36  CLT=C1T+FLM*C2T
                           CS(IS)=CST $ CL(IS)=CLT
      IQ=IS $ CALL EQUIL
  15  I=I-1 $ C2T=ETL(I)*C2T+FTL(I) $ C1T=ET1(I)*C1T+FT1(I)
      C3T=FTS(I)*C3T+FTS(I)
      IF(IZ.EQ.2) GO TO 37 $ CST=C3T+FLN*C2T $ GO TO 38
  37  CST=C3T+FNM*C1T
  38  CLT=C1T+FLM*C2T $ CL(I)=CLT $ CS(I)=CST $ IQ=I $ CALL EQUIL
      IF(I.GT.IMIN) GO TO 15
      FDN=FDN+FLOAT(IT)/FLOAT(IS-IMIN+1)
      RETURN
      END
```

```
      SUBROUTINE EQUIL
      COMMON/FLAG/IZ
      COMMON/EQDATA/WLAST(101),CM(101),CN(101)
      COMMON C1( 101),C2( 101),C3( 101),CL( 101),CS( 101),R( 101),
     1DXL(101),DXS(101)
      COMMON/PARAM/IS,M,N,KM,KN,D1,D2,D3,FSTAR,FLM,FLN,CDUM(6),
     1SPEED,S1,S2,S3,     DELI ,DELR ,RMFN,RBOT,TIME,IMIN,DELT,K1
     2,ESM,ESN,EDN,EPS,   DEL2,DS,WMX,WMN
      COMMON/EQPAR/CLN,CSN,I,IT,REM,FNM
      REAL KM,KN,K1
      IF(CLN.LE.0.0) GO TO 6
      GO TO (14,15),IZ
   14 B=FNM*CLN/CSN $ CST=KN*CSN $ CLT=KM*CLN
      WMAX=1.
      IF(B.GT.1.0) WMAX=1./B
      WMINUS=WMAX
      YMINUS=-WMAX*REM
      YPLUS=-1.0
      WPLUS=0
      WMIN=0
      IF(CL(I).GT.0.0) GO TO 1 $ W=0        $ GO TO  3
    1 W=WLAST(I)+(DXS(I)/B+WLAST(I)/CN(I))*(CSN-CN(I))+(CLN-CM(I))*DXL(I
     1)
    3 A=K1/CLN
      IF(W.GE.WMAX) W=.5*WMAX            $ IF(W.LT.0.0) W=0
    9 W1=1.-W
      W2=1.-B*W
      X=A*((CLT*W1    )**M)*((CST     *W2)**N)
      IT=IT+1
      Y=X-W*REM
      IF(ABS(Y).LE.EPS) GO TO 11
      IF(Y.GT.0.0) GO TO 2
      YMINUS=Y
      WMINUS=W
      GO TO 4
    2 YPLUS=Y
      WPLUS=W
    4                         DY=        FLM/W1+FLN*B/W2
      IF(X.NE.0.0) WMIN=(W*DY+1.0)/(DY+REM/X)
      IF(WMIN.LT.WPLUS) WMIN=WPLUS
      IF(YPLUS.LE.0.0) GO TO 5
      WMAX=WMINUS+YMINUS*(WPLUS-WMINUS)/(YPLUS-YMINUS)
    5 IF(WMAX.GT.WMINUS) WMAX=WMINUS
      W=.5*(WMAX+WMIN)
      IF(WMAX-WMIN.LT.EPS) GO TO 11
      GO TO 9
   11 DY=FLM/W1+FLN*B/W2
   13 A=1.0+W*DY $DXL(I)=W*(FLM-1.0-FLN*B*W/W2)/(A*CLN)
      DXS(I)=W*B*(FLN-1.0-FLN*W/W1)/(A*CSN)
      WLAST(I)=W
      C2(I)=W*CLN*REM
      C1(I)=CLN    -FLM*C2(I)
      C3(I)=CSN    -FLN*C2(I)
      GO TO 7
    6 C2(I)=0 $ C1(I)=CLN        $ C3(I)=CSN $ DXL(I)=0   $ DXS(I)=0
    7 CM(I)=CLN $ CN(I)=CSN
      RETURN
   15 CALL EQUIL2 $ RETURN
      END
```

Appendix: A Computer Code for Sedimentation Calculations

```
      SUBROUTINE EQUIL2
      COMMON/EQDATA/WLAST(101),CM(101),CN(101)
      COMMON C2( 101),C1( 101),C3( 101),CL( 101),CS( 101),R( 101),
     1DXL(101),DXS(101)
      COMMON/PARAM/IS(3) ,KM,KN,D1,D2,D3,ESTAB,REM,FNM,CDUM(6),
     1SPEED,S1,S2,S3,        DEL1 ,DELR ,RMEN,RBOT,TIME,IMIN,DELT,K1
     2,FSM,FSN,FDN,FPS,      DEL2,DS,WMX,WMN
      COMMON/EQPAR/CLN,CSN,I,IT,FLM,FLN
      REAL KM,KN,K1,N,M
      EQUIVALENCE(N,FLN),(M,FLM)
      IF(CLN.LE.0.0) GO TO 6
      CLN=CLN*FLM
      B=FNM*CLN/CSN $ CST=KN*CSN $ CLT=KM*CLN
      WMAX=1.
      IF(B.GT.1.0) WMAX=1./B
      WMINUS=WMAX
      YMINUS=-WMAX*REM
      YPLUS=-1.0
      WPLUS=0
      WMIN=0
      IF(CL(I).GT.0.0) GO TO 1 $ W=0       $ GO TO  3
    1 W=WLAST(I)+(DXS(I)/B+WLAST(I)/CN(I))*(CSN-CN(I))+(CLN-CM(I))*DXL(I
     1)
    3 A=K1/CLN
      IF(W.GE.WMAX) W=.5*WMAX           $ IF(W.LT.0.0) W=0
    9 W1=1.-W
      W2=1.-B*W
      X=A*((CLT*W1    )**M)*((CST    *W2)**N)
      IT=IT+1
      Y=X-W*REM
      IF(ABS(Y).LE.FPS) GO TO 11
      IF(Y.GT.0.0) GO TO 2
      YMINUS=Y
      WMINUS=W
      GO TO 4
    2 YPLUS=Y
      WPLUS=W
    4                       DY=      FLM/W1+FLN*B/W2
      IF(X.NE.0.0) WMAX=(W*DY+1.0)/(DY+REM/X)
      IF(WMAX.GT.WMINUS) WMAX=WMINUS
      IF(YPLUS.LE.0.0) GO TO 5
      WMIN=WMINUS-YMINUS*(WPLUS-WMINUS)/(YPLUS-YMINUS)
    5 IF(WMIN.LT.WPLUS) WMIN=WPLUS
      W=.5*(WMAX+WMIN)
      IF(WMAX-WMIN.LT.FPS) GO TO 11
      GO TO 9
   11 DY=FLM/W1+FLN*B/W2
   13 A=1.0+W*DY $DXL(I)=W*(FLM-1.0-FLN*B*W/W2)/(A*CLN)
      DXS(I)=W*B*(FLN-1.0-FLM*W/W1)/(A*CSN)
      WLAST(I)=W
      C2(I)=W*CLN*REM
      C1(I)=CLN  -FLM*C2(I)
      C3(I)=CSN  -FLN*C2(I)
      GO TO 7
    6 C2(I)=0 $ C1(I)=CLN        $ C3(I)=CSN $ DXL(I)=0   $ DXS(I)=0
    7 CM(I)=CLN $ CN(I)=CSN
      RETURN
      END
```

VI / PRACTICAL IMPLICATIONS

The foregoing chapters have dealt with the theory of mass transport of reacting systems as it applies to the various interactions exhibited by biological macromolecules. Actual experimental systems have been chosen to illustrate the various procedures and precautions required to assure accurate interpretation of sedimentation and electrophoretic patterns in terms of the thermodynamic and molecular parameters characterizing the reactions involved. It seems desirable at this point to summarize these procedures and precautions along with those required for unambiguous interpretation of the more common analytical applications of ultracentrifugation, electrophoresis, and chromatography.

The most dramatic of the new insights provided by these theoretical and experimental investigations is the realization that the electrophoretic, sedimentation, and chromatographic patterns of homogeneous, but interacting, macromolecules can and do show multiple peaks and zones despite instantaneous establishment of equilibrium. This is of great practical significance for it is a disturbing fact that many such patterns could easily be misinterpreted as indicating inherent heterogeneity. It cannot be overemphasized that unequivocal proof of inherent heterogeneity is afforded only by isolation of the various components. Even when an interaction is recognized as such, the patterns can be interpreted in the conventional fashion only when the reactions are sufficiently slow that significant reequilibration does not occur during separation of the various species. When reequilibration occurs between species during their differential transport, peaks or zones in the patterns cannot be placed into correspondence with individual reactants and products, and an entirely different method of analysis of the patterns is required. Moreover, several physicochemical methods must be brought to bear in order to establish the nature of the interaction. For example, the nonenantiographic electrophoretic patterns of β-lactoglobulin (Fig. 36, Chapter III) were recognized by Ogston and Tilley (1) as indicative of an interaction; comparison with sedimentation patterns (Fig. 33, Chapter III) revealed the interaction to be a polymerization of the protein. Subsequent elucidation of this highly interesting tetramerization reaction by Timasheff and his co-workers (2–8) depended on the combined use of a variety of methods including ultracentrifugation, electrophoresis, and light-scattering, low-angle X-ray diffraction, and biochemical techniques for characterization of the three genetic variants of β-lactoglobulin. These investigations illustrate that the greater the diversity of methods brought to bear on the problem, the more precisely can the interaction be described. Furthermore, their success stems in part, from an appreciation of recent developments in the theory and practice of mass transport of interacting systems, whose implications are of such import as to justify further comment.

VELOCITY SEDIMENTATION

For the purpose of discussion, let us consider a hypothetical situation in which a highly purified protein is examined in the analytical ultracentrifuge and found to give velocity sedimentation patterns showing two peaks. The peaks may be rather poorly resolved, as in the case of β-lactoglobulin A (Fig. 36, Chapter III), or very well resolved, as with myosin (Fig. 55, Chapter IV). The questions posed are, first, "Do the two peaks reflect an inherent heterogeneity, or a macromolecular interaction of some sort?" and second, "If they arise from an interaction, what procedures should be followed in order to establish the nature of the interaction?" The following approach, designed to answer these questions, is a synthesis of the various aspects of the mass transport of interacting systems discussed in the preceding chapters. Although each investigator will naturally seek solutions according to his individual style and tastes, successful prosecution requires that cognizance be taken of the several precautions and diagnostic procedures delineated below.

Assuming that after further recrystallizations and/or other purification procedures the protein still shows two sedimenting peaks, fractionation using a partition cell is indicated. In many instances, this will provide an unambiguous answer to the first question. (An exception would be complex formation between two proteins difficult to separate by repeated recrystallization.) The material disappearing across the slower-sedimenting peak is isolated and examined in the ultracentrifuge under the same environmental conditions and at the same protein concentration as used with the unfractionated material. For interactions aside from the exception noted above, the fraction will behave like the unfractionated material and show two peaks, while for inherent heterogeneity a single peak will be obtained. It is conceivable that in the case of an interaction, such an experiment might also yield information on the rate of interconversion of species. Thus, for a sufficiently slow reaction, the faster-sedimenting peak shown by the fraction might grow at the expense of the slower one with time lapsed between fractionation and analysis. Likewise, fractionation on a Sephadex column should help distinguish between heterogeneity and interaction and also between rapid and slow interactions. If the interconversions prove to be slow, conventional analyses like those applied to the slow, reversible dissociation of arachin (*9, 10*) may be used in conjunction with molecular-weight determinations by the Archibald method or by light scattering. In the case of more or less rapid interconversion, the following procedure can be adopted.

A preliminary Archibald molecular-weight determination might be made to decide between macromolecular isomerization and polymerization. Let us suppose that polymerization is indicated. After establishing the optimal conditions of pH, ionic strength, temperature, etc. for polymerization, it is advisable to make an exploratory velocity-sedimentation experiment in a preformed density gradient to detect possible inverted gradients of protein concentration (*11, 12*) arising, for example, from a pressure-sensitive reaction (*11*). In the same vein, sedimentation patterns should be obtained at various rotor speeds. Sensitivity of the sedimentation pattern to rotor speed may be caused by any of several factors: (1) pressure sensitivity of rapidly established equilibria[1] (Fig. 55, Chapter IV); (2) rapidly reequilibrating, ligand-mediated dimerization (Fig. 81, Chapter IV); (3) reaction rates of the order of the rate of separation of interacting species; and (4) kinetically controlled phase changes (*12*). Pressure sensitivity can be distinguished from other causes by adjusting the hydrostatic pressure on the protein solution in the centrifuge cell. This is done by overlaying with increasing thickness of mineral oil at constant rotor speed (Fig. 58, Chapter IV). If ligand-mediated polymerization is suspected, sedimentation experiments can be made with varying concentrations of the ligand. In the case of mediated dimerization, the two peaks of the reaction boundary progressively coalesce with increasing ligand concentration until the boundary becomes unimodal (Fig. 83, Chapter IV).

It now becomes paramount to make sedimentation analyses over a wide range of protein concentration, at low rotor speed if the reaction proves to be pressure sensitive. Experiments at a single concentration or over a limited

---

[1] Even in the case of a homogeneous protein for which no interaction is suspected, it is necessary (*11*) to show that the sedimentation coefficient is independent of rotor speed before computation of molecular weight. This is so because a rapidly reversible, pressure-sensitive dimerization or progressive polymerization reaction, which might occur to a negligible extent at low rotor speeds, could conceivably assume major proportions under the influence of the high pressure generated in the ultracentrifuge cell at high speeds. It is preferable to determine molecular weights by the low-speed Archibald method or by low-speed sedimentation equilibrium.

range of low concentrations could be quite misleading. Consider, for example, a monomeric protein coexisting in rapid equilibrium with a single type of higher-order polymer ($nM \rightleftharpoons P$, $n \geq 3$). At sufficiently low concentration, the sedimentation patterns of this system will show a single peak whose sedimentation coefficient decreases with decreasing concentration and extrapolates to that of the monomer. A limited set of observations of this sort could easily be misinterpreted in terms of a dimerization reaction. Indeed, the association–dissociation reaction might even go unsuspected if only an isolated observation of low concentration were made. On the other hand, if the concentration were to be progressively increased, a second, more rapidly sedimenting peak would eventually appear and grow in area, while that of the slower peak would now remain constant. Constancy of the area of the slower peak with increasing concentration is diagnostic for higher-order polymerization reactions of the type under consideration. Moreover, the sedimentation coefficient of the slower peak in the bimodal reaction boundary will be independent of concentration, while that of the faster one, in general, will show the characteristic concentration dependence illustrated in Fig. 32, Chapter III. Extrapolation of these data to infinite dilution gives the sedimentation coefficient of the monomer. It must be borne in mind, however, that, except for the limiting case of very large n, there is no direct way of determining the sedimentation coefficient of the polymer from the patterns.

Such velocity sedimentation measurements circumscribe the problem, and, in conjunction with ultracentrifugal field relaxation experiments (*13*), may establish the reversibility of the reaction. However, appeal must now be made to independent physicochemical methods. Thus, for example (*4*), the question of reversibility can be examined by light scattering, which also provides crucial information concerning the rates of reaction, degree of polymerization, and equilibrium constant. The degree of polymerization may also be derived from molecular-weight measurements using the Archibald method. With this information in hand, the equilibrium constant for a rapidly reequilibrating polymerization of the type $nM \rightleftharpoons P$ can be computed from the sedimentation patterns with the aid of the Gilbert theory (*14–16*) as described in Chapter III. The value thus obtained is to be compared with that derived from light-scattering measurements.

Many of the same procedures and precautions delineated for association–dissociation reactions can and should be applied to bimolecular complex formation, $A + B \rightleftharpoons C$. Thus, for example, fractionation permits positive identification of peaks in sedimentation (and electrophoretic) patterns as corresponding to pure reactants or constituting reaction boundaries. Also, measures should be taken to ascertain whether the reaction is pressure sensitive; the Gilbert–Jenkins theory (*17*) and its extension (*11*) to include pressure-sensitive systems must serve as interpretative guides. In certain instances, the weak-electrolyte moving-boundary theory can be applied (*18, 19*) to quantitative interpretation of the patterns, as described in Chapter II. In so doing, it is important to show that the value of the equilibrium constant thus obtained is independent of the constituent composition of the initial equilibrium mixture within the range over which a single complex is formed.

Above, we have dwelled on rapidly established equilibria. Kinetically controlled processes having half-times of the order of the time of sedimentation present special interpretative difficulties. The patterns may show two or three peaks which comprise a single reaction boundary and whose sedimentation coefficients bear a complicated relationship to those of the interacting species. Nor are the reaction rates sufficiently slow to justify placing the areas of the peaks into correspondence with the concentrations of individual species. Interactions of this sort may sometimes be recognized by their sensitivity to total macromolecular concentration. Thus, in the case of kinetically controlled dimerization, the theoretical sedimentation (and electrophoretic) patterns are altered in a dramatic fashion when the concentration is changed (*20*). Kinetically controlled dimerization and isomerization reactions can also be detected by the distinctive manner in which the shape of their reaction boundaries changes with time of sedimentation or electrophoresis (Figs. 52–54, Chapter IV).

Finally, it should be noted that certain reversible macromolecular interactions can, in principle, cause multiple zoning in band sedimentation and zone sedimentation in a preformed density gradient. The implications for the many conventional applications of these methods are obvious. Possible complications of this sort will become evident, however, when the fractions themselves are examined in the ultracentrifuge to see if they run true.

ELECTROPHORESIS

Moving-Boundary Electrophoresis. It is well established that the moving-boundary electrophoretic patterns of highly purified proteins may show multiple peaks due to macromolecular interactions despite instantaneous establishment of equilibrium. Examples include the tetramerization of β-lactoglobulin A (Fig. 36, Chapter III) and the interaction of a variety of proteins with undissociated buffer acid (Figs. 68 and 69, Chapter IV). It is characteristic of such interactions that the electrophoretic patterns usually exhibit a nonenantiography which is quite bizarre and not at all like that shown by mixtures of noninteracting proteins (of the above-mentioned figures, compare Figs. 36 and 68 with Fig. 4, Chapter I). Whenever such nonenantiographic patterns are encountered, they must be viewed as suggestive of possible interaction. Cognizance must also be taken of the fact that the patterns may sometimes be reasonably enantiographic, aping a mixture of noninteracting proteins [Fig. 69, Fig. 72 ($n = 100, 300$), and Fig. 75, Chapter IV]. Clearly, conclusions as to inherent heterogeneity ultimately depend upon fractionation experiments.

If certain precautions are taken, fractionation provides an unambiguous method for distinguishing between interaction and inherent heterogeneity. The protein disappearing across the fast-moving ascending and slow-moving descending peaks is isolated. The resulting fractions are then analyzed electrophoretically under conditions identical with those used in the original separation. For interactions, the fractions will behave like the unfractionated material and show multiple peaks, while, for heterogeneity, a single peak will be obtained. It must be stressed that validity of the fractionation test rests upon fulfillment of two requirements. First, the fractions should be analyzed at about the same protein concentration as the unfractionated control. Although the reason for taking this precaution is self-evident in the case of macromolecular association–dissociation, it may not be so apparent for protein–buffer interaction. If the interaction is with the buffer and the concentration of the fraction is too low, the ratio of protein concentration to that of the reacting constituent of the buffer may approach the limit in which the latter is in such large excess that the system behaves like a single component with average transport properties. In that event, the patterns of the fraction will show a single peak, thereby leading to an incorrect conclusion of heterogeneity. It is imperative, therefore, either to concentrate the fraction before dialysis or to calibrate the method by analyzing the unfractionated material at several different concentrations. The second requirement is that, prior to analysis, the fraction be reequilibrated by dialysis against the buffer or readjusted to the same conductance and pH as the buffer by addition of a small quantity of a buffer of appropriate concentration and pH. This is an absolute requirement, for, if it is not met, a fraction from an interacting system of the protein–buffer type will show a single peak, once again leading to an incorrect conclusion of heterogeneity.

Protein–buffer interaction may be distinguished from other types of interaction by the combined use of ultracentrifugation and electrophoresis. After eliminating association–dissociation reactions by sedimentation analyses, protein–buffer interaction can be identified electrophoretically by systematic variation of buffer concentration at constant pH and ionic strength. As illustrated in Figs. 68 and 69 (Chapter IV), increasing the buffer concentration results in progressive and characteristic changes in the patterns, notably in the appearance and growth of the fast-moving peaks at the expense of slow ones. The question naturally arises as to how one can characterize highly purified proteins with respect to parameters such as their isoelectric pH when strong interactions with the solvent is indicated and noninteracting buffers cannot be found. The method suggested by both theory and experimentation would be to lower the protein concentration and/or increase the buffer concentration to the point where the concentration of the interacting constituent of the buffer is in sufficiently large excess as to be effectively constant. Under such conditions, the electrophoretic patterns of a truly homogeneous, but interacting protein will show a single peak.

Zone Electrophoresis. In the past, the assumption had often been made that zone electrophoresis is immune to the several nonideal effects that sometimes complicate moving-boundary electrophoresis. However, as is evident from the theoretical and experimental considerations presented in Chapter IV, the same caution must be exercised in interpreting zone electrophoretic patterns as moving-boundary patterns. In particular, cognizance must be taken of the fact that multiple zones need not necessarily indicate heterogeneity. We have seen how a single macromolecule interacting reversibly with a small, uncharged constituent of the buffer can give two zones despite instantaneous establishment of equilibrium. The situation can be even more complex. Thus, a single macromolecule which isomerizes reversibly at rates comparable to the rate of electrophoretic separation of the isomers can give two or three zones. Bimolecular-complex formation also can give rise to two or three zones even for rapid reequilibration. Situations of this sort might obtain, for example, when two separable proteins must be combined specifically to give an active enzyme. Clearly, conclusions concerning the molecular nature of substances such as enzymes from zone electrophoresis on complex biological tissues and fluids such as serum may be subject to considerable uncertainty unless the possibility of interactions of macromolecules with each other is carefully elucidated.

Although there are only scattered reports in the literature of multiple zones due to protein–buffer interaction, it appears likely that this phenomenon may be of more general occurrence than recognized. Both theory and experiment (Figs. 78 and 79b, Chapter IV) indicate that such patterns could easily be misinterpreted to mean inherent heterogeneity. Here, as in the case of analytical ultracentrifugation and moving-boundary electrophoresis, appeal must be made to fractionation in order to decide whether multiple zones are due to interaction or to heterogeneity. The protein eluted from each unstained zone is subjected to zone electrophoresis in the same buffer that was used for the original separation. For interactions, the fractions will behave like the unfractionated material and show multiple zones (Figs. 79 and 80, Chapter IV), while, for heterogeneity, a single zone will be obtained.

Validity of the fractionation test is critically dependent upon careful scrutiny of the concentration dependence of the electrophoretic patterns. While, in certain instances, the generation of multiple zones due to interaction may be relatively insensitive to protein concentration, in general, this will not be so.

A case in point is the interaction of conalbumin with borate buffer at pH 8.9 (*21*). This system gives a single zone at low protein concentration, two at intermediate concentration, and three at high concentration. Bovine serum albumin exhibits similar properties in borate buffer, pH 7.2 (*22*). Although resolution into two or three zones occurs at protein concentrations in the range 1–10%, patterns obtained at lower concentrations show only a single zone. The possibility of macromolecular association was eliminated in both cases by ultracentrifugal analyses. Experiments on the albumin system designed to distinguish between interaction with the buffer and heterogeneity illustrate the sole pitfall of the fractionation test alluded to above. Thus, a preliminary experiment gave a fraction whose concentration was inadvertently so low that its electrophoretic pattern showed only a single zone as in a control analysis of unfractionated protein at low concentration. In subsequent experiments, therefore, care was taken to obtain fractions of sufficiently high concentration to ensure against possible misinterpretation should they, too, give a single zone. These fractions consistently showed not one, but two or three zones (Fig. 80, Chapter IV), thereby demonstrating that the multiple zones arise from interaction of albumin with borate buffer and are not indicative of inherent heterogeneity. Now, an isolated observation of the sort made in the preliminary experiment, but without the appropriate control, most likely would be misinterpreted to mean heterogeneity. The recommended procedure for dealing with this problem involves preliminary examination of the concentration dependence of the zone patterns of the unfractionated protein, fractionation at a sufficiently high protein concentration to ensure against possible misinterpretation in case the fractions show a single zone on reanalysis, or restoration of the concentration of the fractions to that used in the original separation, followed by dialysis. The requirement that unconcentrated fractions be reequilibrated by dialysis against buffer prior to analysis, while an absolute one in moving-boundary electrophoresis, will, in general, be met by the elution process itself. In some cases, perhaps dependent upon the nature of the supporting material and the mechanism of insertion of the sample, this requirement may not even be critical for zone electrophoresis.

Aside from fundamental investigations on protein interactions *per se*, it is obviously desirable to avoid conditions conducive to the production of electrophoretic patterns which do not faithfully reflect the inherent state of homogeneity of the protein. Such complications of conventional electrophoretic analysis can often be avoided simply by appropriate choice of buffer. Examples of protein–buffer interactions have been presented in Chapter IV for carboxylate, phosphate–borate, borate, and tris–borate buffers. There is also ample evidence for multiple electrophoretic peaks generated by interaction with acidic solvents containing amino acids (*23–26*). Choice of a buffer in which a particular protein at specified pH and ionic strength will not show multiple peaks and zones due to interaction is largely empirical, but there are some guidelines. For example, electrophoresis in phosphate and barbital buffers appears to be free from these complications, at least for bovine serum albumin at pH 8.9, 7.2, 6.1, and 2.3 [compare Figs. 79a and b, Chapter IV, and Figs. 1c and d of Cann (22); also, see Table VI of Wood (25)]. If strong interaction with the solvent is unavoidable, one can attempt to minimize interpretative difficulties by lowering the protein concentration and/or increasing the buffer concentration. At a sufficiently low protein to buffer concentration ratio, a truly homogeneous, but interacting protein will, in principle, show a single zone. However, it is conceivable that, in certain instances, these conditions may be difficult to realize in practice, and failure to do so must not be accepted in lieu of fractionation as indicative of inherent heterogeneity.

CHROMATOGRAPHY

Many of the considerations elaborated above for ultracentrifugation and electrophoresis also apply to chromatography. Thus, it has been recognized for many years that a single pure substance may give two or more zones on a column or spots on a paper strip, and that it is advisable to rechromatograph fractions to see if they run true. Multiple zoning of this sort can be particularly misleading with mixed chromatograms where an unknown and a known compound are mixed and chromatographed; there is a natural tendency to assume that if two spots appear, the compounds are different, which may not be the case at all. The recommended procedure for circumventing this problem is to chromatograph the unknown, the suspected known, and their mixture side by side so that artifactual double spotting, if it occurs, is obvious.

The various causes of multiple zoning by pure substances have been delineated by Keller and Giddings (27), who also discuss methods for determining the actual cause in particular instances. Multizoning may be purely physical in origin, e.g., overloading, presence of contaminants which may act as displacers during development, and discontinuities in the immobile or mobile phase. Pertinent to this discussion is multizoning due to chemical reaction such as complex formation or isomerization. Thus, some solutes may, unknown to the investigator, undergo chemical conversion from one form, or species, to another prior to application to the column or paper; after application, but before development; or during the course of development. In fact, paper chromatography is a useful means for studying the kinetics and mechanisms of certain types of reaction. An example (28) is the reaction of carbohydrates with ammonia to give three chromatographically discernible products. Prior to this investigation, it was common practice to add ammonia to certain developing solvents in the paper-chromatographic analysis of reducing sugars, which procedure resulted in multispotting due to reaction during the course of development.

The phenomenon of multizoning or spotting due to reversible reaction during the course of development of the chromatogram is well established both in theory and practice. Thus, the theory of chromatography of isomerizing systems, $A \rightleftharpoons B$, has been elaborated by Keller and Giddings (27). As in the case of electrophoresis and sedimentation, the chromatogram can show one, two, or three zones or spots, depending upon the rates of interconversion. An example (29) is the interconversion of glucuronic acid and its lactone, glucurone, during the course of development on paper. Complex formation has long been recognised as a source of multizoning, and the theory (30) of bimolecular complex formation $(A + B \rightleftharpoons C)$ in countercurrent distribution is equally applicable to chromatography. Such a reaction can also give rise to one, two, or three zones, depending upon the value of the equilibrium constant and the length of the column. This is so even for very rapid reactions. An example is the chromatography of tryptophan synthetase on DEAE cellulose (31). In this system, two separable proteins must be combined to give an enzyme. The chromatogram shows two zones corresponding to the separated proteins and a third one of intermediate position containing the active complex.

Thus, we conclude by noting the twofold contribution of the theory of mass transport of interacting systems to biology. First, it provides the insights required for unambiguous interpretation of the results of conventional analytical applications of sedimentation, electrophoresis, and chromatography. More importantly, this same understanding is a prerequisite for precise characterization of macromolecular interactions by these methods and suggests new applications to biological systems.

REFERENCES

1. A. G. Ogston and J. M. A. Tilley, *Biochem. J.* **59**, 644 (1955).
2. S. N. Timasheff and R. Townend, *J. Am. Chem. Soc.* **82**, 3157 (1960).
3. R. Townend, R. J. Winterbottom, and S. N. Timasheff, *J. Am. Chem. Soc.* **82**, 3161 (1960).
4. R. Townend and S. N. Timasheff, *J. Am. Chem. Soc.* **82**, 3168 (1960).
5. R. Townend, L. Weinberger, and S. N. Timasheff, *J. Am. Chem. Soc.* **82**, 3175 (1960).
6. S. N. Timasheff and R. Townend, *J. Am. Chem. Soc.* **83**, 464 (1961).
7. T. F. Kumosinski and S. N. Timasheff, *J. Am. Chem. Soc.* **88**, 5635 (1966).
8. J. J. Basch and S. N. Timasheff, *Arch. Biochem. Biophys.* **118**, 37 (1967).
9. P. Johnson and E. M. Shooter, *Biochim. Biophys. Acta* **5**, 361 (1950).
10. P. Johnson, E. M. Shooter, and E. K. Rideal, *Biochim. Biophys. Acta* **5**, 376 (1950).
11. G. Kegeles, L. Rhodes, and J. L. Bethune, *Proc. Natl. Acad. Sci.* **58**, 45 (1967).
12. J. Rosenbloom and V. N. Schumaker, *Biochemistry* **2**, 1206 (1963).
13. G. Kegeles and C. L. Sia, *Biochemistry* **2**, 906 (1963).

14. G. A. Gilbert, *Discussions Faraday Soc.* **20**, 68 (1955).
15. G. A. Gilbert, *Proc. Roy. Soc.* (*London*) **A250**, 377 (1959).
16. G. A. Gilbert, *Proc. Roy. Soc.* (*London*) **A276**, 354 (1963).
17. G. A. Gilbert and R. C. Ll. Jenkins, *Proc. Roy. Soc.* (*London*) **A253**, 420 (1959).
18. L. G. Longsworth, *in* "Electrophoresis Theory, Methods and Applications" (M. Bier, ed.), Chapter 3. Academic Press, New York, 1959.
19. J. R. Cann and J. A. Klapper, Jr., *J. Biol. Chem.* **236**, 2446 (1961).
20. D. F. Oberhauser, J. L. Bethune, and G. Kegeles, *Biochemistry* **4**, 1878 (1965).
21. W. C. Parker and A. G. Bearn, *Nature* **199**, 1184 (1963).
22. J. R. Cann, *Biochemistry* **5**, 1108 (1966).
23. S. J. Singer and D. H. Campbell, *J. Am. Chem. Soc.* **77**, 3504 (1955).
24. R. A. Phelps and J. R. Cann, *J. Am. Chem. Soc.* **79**, 4677 (1957).
25. E. F. Wood, *J. Phys. Chem.* **62**, 308 (1958).
26. H. M. Dintzis, S. N. Timasheff, and S. J. Singer, unpublished work quoted by R. A. Brown and S. N. Timasheff, *in* "Electrophoresis Theory, Methods, and Practice" (M. Bier, ed.), Chapter 8. Academic Press, New York, 1959.
27. R. A. Keller and J. C. Giddings, *J. Chromatog.* **3**, 205 (1960).
28. I. D. Raacke-Fels, *Arch. Biochim. Biophys.* **43**, 289 (1953).
29. S. M. Partridge and R. G. Westall, *Biochem. J.* **42**, 238 (1948).
30. J. L. Bethune and G. Kegeles, *J. Phys. Chem.* **65**, 1755 (1961).
31. I. P. Crawford and C. Yanofsky, *Proc. Natl. Acad. Sci.* **44**, 1161 (1958).

AUTHOR INDEX

Numbers in parentheses are reference numbers and indicate that an author's work is referred to although his name is not cited in the text. Numbers in italics show the page on which the complete reference is listed.

Abramson, H. A., 4 (15), 8, *44*
Ackers, G. K., 123, 125 (45), 127 (45), 128, 129 (45), 130, 149 (45), *150*
Alberty, R. A., 47, 57 (13), 58 (15), 59 (13), 61 (13), 63 (13), 64 (13), 69, *91*
Alexander, A. E., 9 (21), *45*
Ames, B. N., 19 (38), 20 (38), 42 (38), 43 (38), *45*
Ames, W. F., 95, 96 (7), 97, *149*
Andrews, P., 148 (70), *150*
Aoki, K., 123, *150*
Archibald, W. J., 40, *46*
Aten, J. B. T., 197 (48), *206*

Bailey, H. R., 157 (7), 158, *205*
Baine, P., 8 (20), 18 (20), *45*
Baldwin, R. L., 35 (53), *46*
Barlow, G. H., 205 (54), *206*
Basch, J. J., 102 (18), *149*, 225 (8), *233*
Bearn, A. G., 195 (43), *206*, 231 (21), *234*
Beaufay, H., 19 (37), 42 (37), *45*
Belford, G. G., 157 (6), *205*
Belford, R. L., 157 (6), *205*
Berthet, J., 19 (37), 42 (37), *45*
Bethune, J. L., 81 (50), *92*, 110, 111 (28), 112 (28), 115 (28), 116 (28), 117, 123, 143 (28), *149*, *150*, 153 (2), 154, 156, 157 (5), 160 (13), 167 (13), 168 (13), 169 (13), 170 (13), 171, 172 (18), 173, 174, 175 (13), 176 (13), *205*, 226 (11), 227 (11), 228 (20), 233 (30), *233*, *234*

Beychok, S., 61 (29), 63 (29), *91*
Bickel, M. H., 70 (34), *92*
Block, R. J., 12, *45*
Bovet, D., 70 (34), *92*
Boyer, S. H., 195 (44), *206*
Briehl, R. W., 205 (53), *206*
Briggs, D. R., 57 (12), 59, 66, 67, *91*
Brown, D. M., 114 (34), 115 (34), *149*
Brown, R. A., 94 (5), 113, *149*, *150*
Brunner, R., 19 (40), 20 (40), 40 (40), *45*

Campbell, D. H., 137, 138, *150*, 231 (23), *234*
Cann, J. R., 8, 9, 10, 11, 16 (23), 17 (23, 23a, 23b), 24, *45*, 55 (10), 58 (18, 19, 20), 60, 69, 81 (17), 85 (18, 19, 20), 88 (18), *91* (19, 20), *91*, 94, 117 (39), 123 (42, 42a), 133, 145, *149*, *150*, 157 (7), 158, 177, 178, 179, 180 (30, 40), 182 (30), 183, 184, 185, 186, 187 (35, 37, 39), 188, 189 (41), 190 (40), 191, 192 (41), 193, 194, 196 (42, 45), 198, 199, 205 (52), *205*, *206*, 228 (19), 231 (22, 24), 232, *234*
Chance, B., 91 (56, 57), *92*
Changeux, J.-P., 62 (28c), 63 (28a, c), 73 (28a, c, 45, 46), 75, *91*, *92*, 177, *206*
Chanutin, A., 70, 71, 79, *92*
Chiancone, E., 130, 133 (48), 149 (48), *150*
Coddington, A., 177 (25), *206*
Cohen, J. A., 197 (48), *206*
Coleman, J. S., 61 (25), 62 (25), *91*

235

Craig, D., 171 (20), *205*
Craig, L. C., 171 (20), *205*
Crank, J., 96 (8), *149*
Crawford, I. P., 175 (21), *205*, 233 (31), *234*
Cremer, D., 3 (9), *44*

Davidson, N., 8 (20), 18 (20), *45*
deDuve, C., 19 (37), 42 (37), *45*
DeVault, D., 105 (23), 120 (23), 121 (23), 125 (23), *149*
Dintis, H. M., 231 (26), *234*
Dishon, M., 23, 29, *45*, 207, *217*
Dismukes, E. B., 47 (4, 5), *91*
Doherty, D. G., 91 (58), *92*
Dole, V. P., 17 (23c), *45*, 54, 55, *91*
Durrum, E. L., 3 (10), 12 (30), *44*, *45*

Edelstein, S. J., 130, *150*
Ehrenpreis, L., 58 (21), 85 (21), *91*
Emenkal, H., 3 (13), *44*
Eriksson-Quensel, I. B., 102, *149*

Fainer, D. C., 195 (44), *206*
Fairclough, G. F., Jr., 148 (71), *150*
Faxen, H., 29, *45*
Field, E. O., 153, *205*
Fincham, J. R. S., 177 (25, 26), *206*
Fixman, M., 111 (30), *149*
Foster, J. F., 123, *150*
Fredericq, E., 72 (42), *92*
Fruton, J. S., 148 (71), *150*
Fujita, H., 19 (36), 20 (36), 26 (36), 35 (36, 53), *45*, *46*, 111, *149*, 202, *206*

Gerhart, J. C., 63 (28a, b), 73 (28a, b), 74, 75, *91*
Gibbs, J. W., 35 (52), 36, *46*
Giddings, J. C., 157 (8, 9), 158 (8, 9), 159 (8, 9), 171 (9), *205*, 232, 233, *234*
Gilbert, G. A., 81 (48), 87 (48), 89 (48), *92*, 93, 95, 103, 105, 108, 112 (31, 32), 114 (33), 115, 116, 117, 129 (1, 2), 130 (48), 131 (48, 50, 50a), 141, 143, 145, 147, 148, 149 (48, 50, 50a, 76), *149*, *150*, *151*, 227, *234*
Gilbert, L. M., 112 (32), 114 (33), 130 (48), 133 (48, 50a), 145, 147, 148, 149 (48, 50a), *149*, *150*

Gilbert, W., 72, *92*, 177 (28, 29), *206*
Goad, W. B., 58, 81 (17), *91*, 177, 178 (30, 39, 40, 42), 180 (30, 40), 182 (30), 183, 184, 185, 186, 187 (39), 188, 189, 190 (40), 191, 192, 193, 196 (42, 45), 198, 199, *206*
Goldberg, R. J., 29, 35 (48), *45*, 137, *150*
Gordon, J. A., 205 (51), *206*
Gorin, M. H., 4 (15), 8 (15), *44*
Gosting, L. J., 81 (49), *92*
Grassman, W., 3 (14), *44*
Grillo, P. J., 117, *150*
Gross, D., 102, *149*

Hannig, K., 3 (14), *44*
Harrington, W. F., 81 (52), *92*, 103 (21), 131 (21), *149*, 161, 162, 165, 166, 169 (17), *205*
Harris, H., 13, *45*
Harting, J., 71 (37), 79 (37), *92*
Hartley, B. S., 103 (21), 131 (21), *149*
Haupt, H., 24 (43), *45*
Hayes, J. E., Jr., 71 (37, 38), 79 (37, 38), *92*
Hearst, J., 197 (46), *206*
Heide, K., 24 (43), *45*
Heidelberger, M., 134, *150*
Heimburger, N., 24 (43), *45*
Henry, D. C., 6, *45*
Hess, E. L., 111 (28), 112 (28), 115 (28), 116 (28), 143 (28), *149*, 153 (2), *205*
Hudson, B. W., 139 (62), *150*
Hughes, W. L., Jr., 72 (43, 44), *92*, 114 (35), *150*

Ilten, D., 144 (68), 148 (68), *150*
Imanishi, A., 205 (50), *206*
Isemura, T., 205 (50), *206*
Isliker, H. C., 133, *150*

Jacob, F., 73 (45), 75, *92*
Jacobsen, C. F., 58 (23), *91*, 178, *206*
Jenkins, R. C. Ll., 81 (48), 87 (48), 89 (48), *92*, 93, 95, 141, 143, 145, *149*, 227, *234*
Johnson, P., 9 (21), *45*, 103 (19, 20), *149*, 226 (9, 10), *233*
Johnston, J. P., 23, *45*
Josephs, R., 81 (52), *92*, 161, 162, 165, 166, 169, *205*

Author Index

Kabat, E. A., 133, *150*
Kakiuchi, K., 205 (50), *206*
Kaplan, N. O., 177 (24), *206*
Karush, F., 62 (26), *91*
Kato, S., 205 (50), *206*
Kauzmann, W., 81 (51), *92*, 176 (22), *206*
Kegeles, G., 81 (50), *92*, 111 (28), 112 (28), 115 (28), 116 (28), 117, 123, 143 (28), *149*, *150*, 153 (2, 4), 154 (4), 156, 157 (5), 160, 167, 168, 169, 170 (13), 171, 172 (18), 173, 174, 175, 176 (13), *205*, 226 (11), 227 (11, 13), 228 (20), 233 (30), *233*, *234*
Keeler, R. A., 157 (9), 158 (9), 159 (9), 171 (9), *205*, 232, 233, *234*
Keilin, K., 91 (54), *92*
Kellett, G. L., 130 (48), 133 (48), 149 (48), *150*
Kendrew, J. C., 19, *45*
Kent, R., 19 (40), 20 (40), 40 (40), *45*
Kimmel, J. R., 114 (34), 115 (34), *149*
Kirkwood, J. G., 50 (6), 58 (22), 85 (22), *91*, 94 (5), *149*
Kirshner, A. G., 111 (29), *149*
Kitto, G. B., 177 (24), *206*
Klapper, J. A., Jr., 58 (18, 20), 85 (18, 20), 88 (18), 91 (20), *91*, 145, *150*, 228 (19), *234*
Klapper, M. H., 205, *206*
Klinkenberg, A., 157 (10, 11), 158 (10, 11), 159 (10, 11), *205*
Klotz, I. M., 59, 61 (24), 62, 68, 71, 79, 80, 81 (53), 85, *91*, *92*, 205 (54, 55), *206*
Kossman, R. J., 195 (44), *206*
Kraus, K., 3 (11), *44*
Kumosinski, T. F., 102 (17), 113 (17), 114 (17), 115 (17), 116 (17), *149*, 225 (7), *233*
Kunkel, H. G., 18, *45*

Lamm, O., 26 (45), 29 (47), *45*
Linder, S. E., 3, *44*
Lodge, O., 3, *44*
Loke, J. P., 131 (49a), *150*
Longsworth, L. G., 1, 3, 16 (22), 17 (22), *44*, *45*, 52 (7), 55 (7), 57, 58 (7, 23), 59, 85 (7), *91*, 121, 146, *150*, 178, *206*, 228 (18), *234*
Loveless, M. H., 11, *45*
Luck, J. M., 62 (28), *91*
Ludewig, S., 70 (35), 71, 79 (35), *92*

McDonald, H., 3 (12), *44*
McDougall, E. I., 70 (33), *92*
MacInnes, D. A., 1, 3, *44*
Mann, T., 91 (54), *92*
Margenau, H., 36 (54), *46*
Martin, R. G., 19 (38), 20 (38), 42 (38), 43 (38), *45*
Marvin, H. H., Jr., 57 (13), 58 (15), 59 (13), 61 (13), 63 (13), 64 (13), 69, *91*
Masket, A. V., 70 (35), 71, 79 (35), *92*
Massey, V., 103 (21), 131 (21), *149*
Mayer, M. N., 133, *150*
Meselson, M., 19 (39), 20 (39), 42 (39), *45*
Monod, J., 73 (45, 46), 75, *92*, 177, *206*
Morton, K. W., 213, 214, *217*
Moyer, L. S., 4 (15), 8 (15), *44*
Müller-Hill, B., 72, *92*, 177 (28, 29), *206*
Muller, H., 4 (16), *44*
Murata, M., 70 (32), *92*
Murphy, G. M., 36 (54), *46*

Nakamura, S., 70 (32), *92*
Neurath, H., 72 (42), *92*
Nichol, J. C., 47 (2, 5), *91*
Nichol, L. W., 110, 111 (28), 112, 115 (28), 116, 131 (49a), 143, 149 (72, 73, 74, 75, 78, 79), *149*, *150*, *151*, 153 (2), *205*
Nisonoff, A., 62 (27), *91*
Novick, A., 177 (27), *206*

Oberhauser, D. F., 157 (5), *205*, 228 (20), *234*
Ogston, A. G., 23, *45*, 102, 149 (75, 78, 79), *149*, 151, 153, *205*, 224, *233*
Olivera, B. M., 8, 18, *45*
Ottesen, M., 205 (51), *206*
Overbeek, J. Th. G., 4 (17, 18), 7 (17, 18), 9 (18), *44*, *45*

Parker, W. C., 195 (43), *206*, 231 (21), *234*
Partridge, S. M., 233 (29), *234*
Pauling, L., 140, *150*
Pederson, K. O., 19 (32), 20, 35 (32), 39 (32), *45*, 134, *150*
Pepe, F. A., 138, 144 (68), 145, 146, 148 (61, 68), *150*
Phelps, R. A., 69, *91*, 117 (39), 123 (42), *150*, 178 (32, 33, 36), 179, 205 (52), *206*, 231 (24), *234*

Picton, H., 3, *44*
Pivan, R. B., 59 (24), 61 (24), 62 (24), 68 (24), 79 (24), 80 (24), 85 (24), *91*
Pressman, D., 62 (27), *91*, 140 (64), *150*

Raacke-Fels, I. D., 232 (28), *234*
Rao, M. S. N., 117, *150*, 153 (4), 154 (4), 205
Reuss, F. F., 3, *44*
Rhodes, L., 81 (50), *92*, 160 (13), 167 (13), 168 (13), 169 (13), 170 (13), 175 (13), 176 (13), *205*, 226 (11), 227 (11), *233*
Richards, E. G., 34 (49), *45*
Richtmyer, R. D., 213, 214, *217*
Rideal, E. K., 103 (20), *149*, 226 (10), *233*
Roark, D., 34 (50), *46*
Rosenberg, R. M., 81 (53), *92*
Rosenbloom, J., 197 (47), *206*, 226 (12), *233*
Ross, P. D., 8 (19), *45*
Rubin, M. M., 62 (28c), 63 (28c), 73 (28c), *91*

Sadler, J. R., 177 (27), *206*
Sanger, F., 19, *45*
Sasaki, I., 70 (32), *92*
Scatchard, G., 61 (25), 62 (25), *91*
Schachman, H. K., 19 (35), 20 (35), 22, 23, 34 (35, 49, 51), 35, 41, *45*, *46*, 57 (14), 63 (28a, b), 73 (28a, b), 74, 75, 76, 80, 83, 84, *91*, *92*, 130, *150*, 168, 169 (16, 17), *205*
Scheraga, H. A., 129 (46), 130, 131 (49), 132, 149 (46, 49), *150*
Schilling, K., 18 (28), *45*, 58 (16), 59 (16), 69, *91*
Scholten, P. C., 157 (12), 159, *205*
Schultze, H. E., 24, *45*
Schumaker, V. N., 197 (47), *206*, 226 (12), *233*
Shen, A. L., 61 (25), 62 (25), *91*
Shooter, E. M., 103 (19, 20), *149*, 226 (9, 10), *233*
Sia, C. L., 227 (13), *233*
Singer, S. J., 133, 135, 137, 138, 139, 144 (68), 145, 146, 148 (61), *150*, 231 (23, 26), *234*
Smith, E. L., 114 (34), 115, *149*
Smith, G., 3 (11), *44*
Smith, R. F., 57 (12), 59, 66, 67, *91*
Stahl, F. W., 19 (39), 20 (39), 42 (39), *45*

Steinberg, I. Z., 34 (51), 35, *46*, 57 (14), 76, 80, 83, 84, *91*, 92
Steinhardt, J., 61 (29), 63 (29), *91*
Stern, K. G., 91 (55), *92*
Svedberg, T., 19 (32), 20, 35 (32), 39 (32), *45*, 102, *149*
Svensson, H., 47, 53, 55 (9), *91*
Synge, R. L. M., 70 (33), *92*

Takeo, K., 70 (32), *92*
Tanford, C., 111 (29), *149*
Teller, D. C., 34 (49), *45*
Ten Eyck, L. F., 81 (51), *92*, 176 (22), *206*
Teresi, J. D., 62 (28), *91*
Thomas, H. C., 105 (25), 120 (25), 121 (25), *149*
Thompson, T. E., 123, 125 (45), 127 (45), 128, 129 (45), 130, 149 (45), *150*
Tilley, J. M. A., 102, *149*, 224, *233*
Timasheff, S. N., 58 (22), 85 (22), *91*, 102, 111 (15), 112 (15), 113 (13, 14, 17), 114 (13, 14, 17), 115 (17), 116 (17), 117 (13), 119 (13), *149*, *150*, 225, 227 (4), 231 (26), *233*, *234*
Tiselius, A., 1, 3 (9), 18, *44*, *45*, 50, *91*, 102, 133 (51), *149*, *150*
Tombs, M. P., 114 (40), 118, 119, *150*
Townend, R., 102 (12, 13, 14, 15, 16), 111 (15), 112 (15), 113 (13, 14), 114 (13, 14), 117 (13), 119, *149*, 225 (2, 3, 4, 5, 6), 227 (4), *233*
Turba, F., 3 (13), *44*

Urbin, M., 3 (12), *44*

Van Holde, K. E., 35 (53), *46*, 94, *149*
Vaslow, F., 91 (58), *92*
Velick, S. F., 71, 79, *92*
Vinograd, J., 19 (39, 40), 20 (39, 40), 40 (40), 42 (39), *45*, 197 (46), *206*

Waldmann-Meyer, H., 18, *45*
Walker, F. M., 59 (24), 61 (24), 62 (24), 68 (24), 79 (24), 80 (24), 85 (24), *91*
Warner, R. C., 58 (21), 85 (21), *91*
Wasserman, P. M., 177 (24), *206*
Waugh, D. F., 144, *150*
Weigle, J., 19 (40), 20 (40), 40 (40), *45*

Author Index

Weinberger, L., 102 (15), 111 (15), 112 (15), *149*, 225 (5), *233*
Weiss, G. H., 23 (42), 29 (42), *45*, 207 (1), *217*
Weiss, J., 105 (24), 120 (24), 121 (24), *149*
Westall, R. G., 233 (29), *234*
Wiersema, P. H., 4 (18), 7 (18), 9 (18), *45*
Williams, J. W., 35 (53), *46*
Williamson, M., 3 (12), *44*
Wilson, J. N., 105, *149*
Winterbottom, R. J., 102 (13), 113 (13), 114 (13), 117 (13), 119 (13), *149*, 225 (3), *233*

Winzor, D. J., 129 (46), 130, 131 (49, 49a), 132, 149 (46, 49, 72, 73, 74, 77), *150, 151*
Wood, E. F., 231 (25), 232, *234*
Wyman, J., 73 (46), *92*, 177, *206*

Yanofsky, C., 175 (21), *205*, 233 (31), *234*
Yphantis, D. A., 19 (41), 20 (41), 23 (42), 29 (42), 34 (50), *45, 46*, 144, *150*, 207 (1), *217*

Zweig, G., 12 (30), *45*

SUBJECT INDEX

A

Allosteric interaction, 73, 177
Ampholyte, *see* Constituent concentration, Constituent mobility, Dissociation equilibria, Electrophoresis, Electrophoretic pattern, Transport-interconversion equation
α-Amylase subunits, 205
Antibodies, 62, 133
  heterogeneity of binding sites, 62, 139
  valence, 134, 137
Antigen–antibody complexes, 134–137
  electrophoresis, 136–137, 148
  ultracentrifugation, 134–136
Antigen–antibody reaction, 133–140
  electrophoresis, 136, 137, 148
  extended Gilbert–Jenkins theory, 148
  forces, 139, 140
  framework theory, 134, 136
  Gilbert–Jenkins theory, 144, 145
  Goldberg theory, 137
  sedimentation, 144, 145
  soluble complex, 134
  thermodynamics, 137–140
  ultracentrifugation, 134–136
Arachin, dissociation of, 103, 226
Archibald method, 34, 113, 153, 171, 226, 227
  principle of, 40
Aspartate transcarbamylase
  allosteric interactions, 73–76
  catalytic subunits, 75
  effect of ligands on morphology, 73
  feedback inhibition, 73
  regulatory subunits, 75

B

Band sedimentation
  interacting systems, 175–176
  kinetically controlled isomerization, 159
  multiple zoning, 228
  pressure-sensitive interactions, 175, 176
  principle of, 40
Boltzmann transformation, 96
Boundary conditions, 211, 212

C

Casein, interaction with Ca^{2+}, sedimentation, 71
Chromatography
  kinetically controlled isomerization, 159
  macromolecular association–dissociation, 171–173
  macromolecular complexes, 173–175, 233
  macromolecular isomerization, 233
  multiple zoning, causes, 232, 233
  precautions, 232
  tryptophan synthetase, 233
α-Chrymotrypsin
  association–dissociation, 131, 153–154
  molecular sieve chromatography, 131
  polymerization, 103
  progressive polymerization, 116, 155
Computational method, summary, 214–215
Computer, 207
Computer code, 207, 217–223
Conalbumin
  complex with lysozyme, 58
  interaction with borate buffer, 195
Conservation equation
  for ampholytes
    applications of integral form of, 52–58
    electrophoresis, 51

Conservation equation—*contd.*
  analytical solution of approximate, 83–95
  comparison of analytical and numerical treatment, 155
  definition of, for electrophoresis, 13
  derivation of
    for electrophoresis, 13–15
    for sedimentation, 26–27
  for electrophoresis
    of ampholytes, 51
    ideal, 15
    macromolecular complexes, 93
    macromolecular isomerization, 49
    polymerization, exact, 155
  sedimentation
    of mixture of macromolecules, 28
    rectilinear and constant field approximation, 28
    velocity sedimentation, 26
  exact, 152
  Gilbert theory, 104
  ideal transport ignoring diffusion
    similarity solution, 101–102
    solution by transformation of position variable, 101
  integral form, 93
  ligand-mediated association–dissociation, numerical solution, 196–204
  macromolecular complexes, 141
  macromolecule–neutral molecule interaction, 181
    numerical solution, 182
  numerical solution of
    methods, 207–223
    results, 152–206
  relations derived therefrom, 29–34
  solution for ideal electrophoresis, 16

Constituent concentration
  of ampholytes, 50–51
    protein and interacting ion, 62
  definition of, 49–50

Constituent diffusion coefficient, definition of, 50

Constituent mobility
  of ampholytes, 51
    homogeneous protein, 54
  definition of, 50
  determination of, 177
  in enzyme–substrate mixture, 86
  protein–small-ion interactions, 63
  from reaction boundary, 64, 177

Constituent sedimentation coefficient
  definition of, 76
  protein–small-ion interaction, 81–82

Continuity equation, *see* Conservation equation, Forced-diffusion equation

Convection
  hypersharpening, 168
  moving-boundary electrophoresis, 186–188
  velocity sedimentation, 168, 170

Coordinates
  discrete, 208
  moving, 214

Countercurrent distribution
  interactions, 155
  macromolecular association–dissociation, 171–173
  macromolecular complexes, 173–175

Crossing-paper electrophoresis, 69–70

D

Density-gradient equilibrium sedimentation, principle of, 41

Difference equations, 209
  implicit, 213

Diffusion coefficient, 5, 14, 23

Dissociation equilibria for ampholytes, 50

DNA
  density-gradient equilibrium sedimentation, 42
  mobility and charge of, 8–9
  skewing and self-sharpening of sedimenting boundary, 23
  zone electrophoresis in density gradient, 18

Dole's strong-electrolyte moving-boundary theory, 17, 54–55

E

Einstein relationship, 5

Electrochromatography, 19

Electromigration
  electrophoretic effect, 5
  governing factors in free solution, 8–10
  on solid supports, 18
  Henry's equation, 6
  Henry–Debye–Hückel equation, 6–7
  relaxation effect, 6
  theory of, 4–10

Electrophoresis
  antigen–antibody reaction, 136–137

Subject Index 243

Electrophoresis—contd.
  combined use with hydrodynamic methods, 10
  definition of, 3
  Gilbert theory, 117–123
  high-voltage, 19
  historical development of, 3
  hypersharp reaction boundary, 122
  ideal moving-boundary, 15–16
  macromolecule–neutral-molecule interaction, 178–179
  phenomenological theory of, 10–19
  principles for ampholytes, 47
  protein–small-ion interactions, 63–70
  relative advantages of moving boundary and zone, 3–4
  serum albumin–methyl orange, 66–69
Electrophoretic charge
  evaluation of, 7–9
  versus titration charge, 8–10
Electrophoretic mobility
  calculation from moving-boundary patterns, 16–17
  definition of, 4
  dependence on electrical charge and frictional coefficient, 4–10
    on net charge, frictional coefficient, ionic strength, buffer composition, 8–10
  Henry's equation, 6
  Henry–Debye–Hückel equation, 6–7
  Influence of small ion binding on, 58
  relationship for nonconducting medium, 5
Electrophoretic pattern
  antigen–antibody, 148
  apparent and actual compositions, moving-boundary, 16–17
  calculation of mobilities from moving-boundary, 16–17
  classical interpretation of, for moving-boundary, 16
  definition of, 12
  enzyme–substrate, quantitative interpretation, 87
  ideal moving-boundary, 16
  interpretation for fast interaction, 103
    for slow interaction, 103
  multiple peaks and zones due to interaction, 93–94
  nonenantiographic, moving-boundary, 16–17

  nonideal moving-boundary, 16–17
    of ampholytes, 47
    analogy with rapid isomerization, 52
    ideal, 52
  pepsin–serum albumin, 146–147
    interpretation in terms of equilibrium constant, 89–90
    nature of, 93
  stationary boundaries, 16–17
Enzyme–protein substrate interaction, electrophoresis, theory, and methods, 85–91
Equilibrium dialysis
  allosteric interactions of aspartate transcarbamylase, 73
  method, 62
  serum albumin–methyl orange, 59, 62, 66–68, 79, 84–85
Equilibrium sedimentation in density gradient, principle of, 42

F

Feedback inhibition, 73
Fick's first law, 13, 26
Fick's second law, 13, 15, 96
  Boltzmann transformation, 99
  for sector-shaped cell, 27
  similarity transformation of, 98–100
  solution by similarity transformation, 98–100
First moment of moving boundary, 52–54, 120–121
  relationship to constituent mobility, 54, 56–57, 121
Forced diffusion, 15, 25
  comparison of electrophoresis and ultracentrifugation, 57
  definition of, 13
Forced diffusion equation
  derivation of
    for electrophoresis, 13–15
    sedimentation, 26–27
  for electrophoresis, 13
    ideal electrophoresis, 15
    rapid isomerization, 49
    velocity sedimentation, 26
  relations derived therefrom, 29–34
  solution for ideal electrophoresis, 16
FORTRAN, 217
Fractionation
  precautions, 187, 195, 229, 230–231

Fractionation—*contd.*
    test for heterogeneity, 171, 225, 229, 230–231
    for interaction, 171, 189, 225, 229
Free diffusion, spreading of boundaries and zones due to, 13
Frictional coefficient
    definition of, 4
    effect of ion binding on, 9–10

G

Gel filtration, *see* Molecular sieve chromatography
Gilbert theory, 95, 102–133, 197, 199, 227
    association of myosin, 161
    electrophoresis, 117–123
    extended, 154
    formulation of, 103–111
    Gilbert's continuous solution, 105–106
    limiting case, 161–164
    progressive polymerization, 117
    sedimentation, 106–117
    similarity transformation, 104
    solution, any initial conditions, 120
    Wilson's discontinuous solution, 105
Gilbert–Jenkins theory, 95, 133–149, 227
    applications, 143–149
    formulation, 141–143
    predictions, 143–144
    similarity solutions, 142
γ-Globulin, 133, 134, 136
    interaction with acetic acid, 182
        with salt ions, 69
    paper electrophoresis of, 18
Glutamic acid dehydrogenase, activation of, 177
Glyceraldehyde-3-phosphate dehydrogenase, interaction with diphosphopyridine nucleotide, sedimentation, 171
Goldberg theory, 137

H

Hemerythrin, dissociation by thiocyanate, 205
Hemoglobin
    dissociation into subunits, 129–130
    molecular sieve chromatography, 129–130, 133
Henry–Debye–Hückel equation, 6–7
Henry's equation, 6
Heterogeneity
    proof of, 224
    tests for, 187, 190, 193–195, 225, 229, 230–231
Hypersharpening of boundary
    due to convection, 168
    reaction, 119–120, 121–122, 129

I

Immunoelectrophoresis, 19
Implicit difference equation, 213
Insulin
    complex with protamine, 58
    interaction with thiocyanate, 72
Interacting systems
    aggregation of proteins by Al^{3+}, 205
    allostery, 73, 177
    α-amylase, 205
    antigen–antibody reaction, 133–140
    aspartate transcarbamylase ligands, 73–76
    association of α-chymotrypsin, 131, 153–155
        of myosin, 161–167
    association–dissociation, 102
    casein–Ca^{2+}, 71
    conalbumin–borate buffer, 195
    conalbumin–lysozyme, 58
    between different macromolecules, 85–91
    dimerization, pressure-sensitive, 167–169, 175–176
    dissociation
        of arachin, 103, 226
        of hemerythrin, 205
        of hemoglobin, 129–130, 133
        of β-lactoglobulin, 103
        of lamprey hemoglobin, 205
    effect of pressure
        on band sedimentation, 175, 176
        theory, 167–170, 175–176
        on velocity sedimentation, 81, 160–171
    enzyme–effector, 57
    feedback, 73
    γ-globulin–acetic acid, 182
    γ-globulin–salt ions, 69
    glutamic acid dehydrogenase, activation of, 177
    glyceraldehyde-3-phosphate dehydrogenase with diphosphopyridine nucleotide, 71
    insulin–protamine, 58
    insulin with thiocyanate, 72
    kinetically controlled, 94, 156–159, 228
    rotor speed, 170

Subject Index

Interacting systems—*contd.*
  β-lactoglobulin–acetate buffer, 189
  ligand-mediated dimerization, 197–202*
  ligand-mediated dissociation, 204
  ligand-mediated polymerization, 226
  ligand-mediated tetramerization, 202–203
  macromolecular association–dissociation, 171–173
  macromolecular complexes, 133–149, 173–176, 233
    pressure-sensitive, 169–170, 175–176
  macromolecular isomerization, 48–50, 94, 157–160, 233
  macromolecule–small-molecule, 176–205, *see also* particular system
  mercaptalbumin with mercuric ion, 72
  numerical treatment of, 182, 207–223
  ovalbumin–nucleic acid, 58
  ovomucoid–RNA, 147–148
  pepsin–serum albumin, 58, 88–91, 146–147
  polymerization of α-chymotrypsin, 103
  practical implications, 224–234
  precautions, 170, 224–234
  progressive polymerization, 154–155
  progressive trimerization, 155
  protein–buffer interaction, 178–190, 229
  protein–neutral molecule, 58
  protein–small ions, 58–85, 177
  protein–undissociated buffer acid, 179
  repressor–inducer–*lac* operator, 72, 177
  serum albumin–amines, 70
  serum albumin–borate buffer, 189, 192–195, 231
  serum albumin–Cd^{2+}, 58, 69
  serum albumin–chloride ion, 57–59, 69
  serum albumin–fatty acid, 62
  serum albumin–methyl orange, 57, 66–69, 79–81
  tests for, 102–103, 109–110, 114–115, 122–123, 144, 157, 179, 187, 190, 193–195, 225, 229, 230–231
    pressure effects, 164–166, 170
  tetramerization of β-lactoglobulin, 102–103, 113–114, 116, 117, 122
  tryptophan synthetase, 175, 233
  zone transport, 171–176
Interaction(s)
  between different macromolecules, 85–91
  numerical treatment of, 182, 207–223
Intrinsic equilibrium constant, definition of, 61

Isozymes, 177

J

Johnston–Ogston effect
  analogy with nonideal electrophoresis, 57
  definition of, 23
  explanation of, 32–34

K

Kinetically controlled interactions
  reaction boundary, 228
  test for, 228
  theory, 94, 156–159, 170

L

β-Lactoglobulin, 224–225
  dissociation into subunits, 103, 111–113
  genetic variance, 103
  interacting with acetate buffer, 189
  interplay between association and hydrodynamics, 115
  progressive polymerization, 116
  tetramerization, 103, 113–114, 117, 122
Lamm equation
  derivation of, 26–27
  relations derived therefrom, 29–34
Lamprey hemoglobin, dissociation by oxygen, 205
Langmuir adsorption isotherm, 59
Ligand-mediated association–dissociation, theory of sedimentation, 196–204, *see also* Interacting systems
Ligand-mediated dimerization, 197–202, *see also* Interacting systems
  effect of centrifugal field, 197
Ligand-mediated dissociation, theory of sedimentation, 204, *see also* Interacting systems
Ligand-mediated polymerization, detection, 226, *see also* Interacting systems
Ligand-mediated tetramerization, theory of sedimentation, 202–203, *see also* Interacting systems
Lysozyme, complex with conalbumin, 58

M

Macromolecular association–dissociation
  chromatography, 171–173
  countercurrent distribution, 171–173
  determination of association constant, 111
  electrophoresis, 102, 117

Macromolecular association–dissociation
—*contd.*
  Gilbert theory, sedimentation, 106–117
  ligand-mediated, 196–205
  molecular sieve chromatography, theory, 127–129
  progressive polymerization, 116–117
  sedimentation, 102, 111–117
  sedimentation coefficients versus concentration, 111–113
  tests for, 227
  zone electrophoresis, 171–173
Macromolecular complexes
  band sedimentation, pressure, 175–176
  chromatography, 173–175, 233
  countercurrent distribution, 173–175
  effect of pressure, 169–170
  Gilbert–Jenkins theory, 143–149
  procedures and precautions, 227
  zone electrophoresis, 173–175
Macromolecular dimerization
  band sedimentation, pressure-sensitive, 175–176
  effect of pressure, 167–169
  kinetically controlled, 156–157
  ligand-mediated, 197–202
Macromolecular interactions, *see also* Interacting systems, specific type of interaction
  numerical treatment of, 182, 207–223
Macromolecular isomerization
  analytical solution of transport-interconversion equations, 94
  band sedimentation, 159
  chromatography, 159, 233
  conservation equation, 49
  exact transport-interconversion equation, 157
  kinetically controlled, 157–159
  moving-boundary electrophoresis, 157–159
  rates of, and electrophoretic pattern, 48–49
  transport-interconversion equation, 48
  velocity sedimentation, 159–160
  zone electrophoresis, 159
  sedimentation, 159–160
Macromolecule–neutral-molecule interaction, 176–205
  electrophoresis, 178–179
  formulation of theory, 180–182

Mass action relationship, protein–small-ion interaction, 61–62
Mercaptalbumin, interaction with mercuric ion, 72
Mercuripapain, sedimentation behavior, 114
Michaelis–Menten complex
  electrophoresis, 146–147
  Gilbert–Jenkins theory, 146–147
  pepsin–serum albumin, 58, 85–91, 146–147
    effect of acetyl-L-tryptophan and fatty acids, 90–91
Michaelis–Menten constant, pepsin–serum albumin, 90
Molecular sieve chromatography (gel filtration)
  association–dissociation, theory, 127–129
  α-chymotrypsin, 131
  description of, 123–125
  differential methods, 131–133
  Gilbert–Jenkins theory, 148–149
  hemoglobin, 129-130, 133
  macromolecular interactions, 148–149
  mathematical theory, 125–127
Molecular sieve coefficient definition of, 125
Moving-boundary electrophoresis
  classical interpretation of, 16
  convection, 186–188
  definition of, 10
  fractionation, 229
  ideal, 16
  kinetically controlled isomerization, 157–159
  multiple peaks due to interactions, 228–230
  nonideal, 16–17
  precautions, 229
  protein–buffer interaction, theory, 182–190
  protein–small-ion interactions, experimental design, 64–66
  serum albumin–Cl^-, 58–59
  serum albumin–methyl orange, 59
  tests for heterogeneity, 229
Moving-boundary equation
  for electrophoresis, 55–57
  sedimentation, 32–33, 78
  practical significance of, 57
  prediction of Johnston–Ogston effect, 33–34

Subject Index

Moving electrophoretic boundary
  first moment of, and constituent mobility, 54, 121
  spreading of, 54–55
Multiple peaks and zones due to interactions, 224, 228, 230
Myoglobin, starch-gel electrophoresis, 195
Myosin
  association, effect of pressure, 161–167
  effect of rotor speed, 164
  velocity sedimentation, effect of rotor speed, 164

N

Newton's method, 215
Nucleic acid, complex with ovalbumin, 58
Numerical methods, 207–223
  for solution of equilibrium condition, 215
Numerical stability, 210, 212

O

Ovalbumin
  complex with nucleic acid, 58
  interaction with buffer acid, 182
Ovomucoid, interaction with RNA, 147–148

P

Pepsin, interaction with serum albumin, 58, 88–91, 93, 146–147
Photoelectric scanner, 72–73
  sedimentation of aspartate transcarbamylase–PMB, 75
  serum albumin–methyl orange, 78–79
Pressure
  band sedimentation, 175–176
  sedimentation equilibrium, 39
  velocity sedimentation, 81, 160–171
  theory, 167–170
  zone sedimentation, 175–176
Pressure-sensitive reaction, detection, 226
Progressive polymerization, 116–117, 154–155
Progressive trimerization, 155
Protamine, complex with insulin, 58
Protein(s)
  aggregation by Al^{3+}, 205
  anion binding and aggregation, 205
  number of ionic species in solution, 58
  subunit structure of, 19
Protein–buffer interaction
  detection, 229
  implications of theory, 190
  moving-boundary electrophoresis, 182–190
  theoretical predictions, 182–190
  zone electrophoresis, 190–196
Protein–neutral-molecule interaction, implications of theory, 190
Protein–small-ion interactions
  constituent mobilities, 63
  electrophoretic measurements and theory, 63–70
  mass action, 59, 61–62
  methods of study, 63
  sedimentation equilibrium, 82–85
  sedimentation measurements, 70–76
  velocity sedimentation, theory, 76–79
Protein–undissociated buffer acid interaction
  formulation of theory, 180–182
  moving-boundary electrophoresis, 178–190
  theoretical predictions, 182–190
  zone electrophoresis, 190–196

R

Radial dilution, 23, 81
  derivation of correction factor, 30–31
Reaction boundary
  bimodal, 144
    Gilbert theory, 109
  constituent mobility, 64
  definition of, electrophoresis, 64
  sedimentation, 77–78
  diffusion, influence on, 153, 156–157
  of enzyme–substrate mixture, 86
  hypersharp, 122
  nature of, 152
  pepsin–serum albumin, bimodality of ascending, 89
  precise shape of, 152–171
  protein–buffer acid interaction, 184
  rate of sedimentation, 110
  shape of, 93, 202
Reaction zone, definition of, 171
Regulatory enzymes
  aspartate transcarbamylase, 73–76
  cooperative binding of substrate, 63
Repressor–inducer–*lac* operator, 72, 177
RNA, interaction with ovomucoid, 147–148
Rotor speed, sensitivity of sedimentation pattern to, 226

S

Sedimentation
  Gilbert theory, 106–117
  theory of, 20–25
Sedimentation coefficient
  association–dissociation reaction, 111–113
  computation of, 21
  from second moment of boundary, 29–30
  definition of, 20
  interpretation of, 22–23
Sedimentation equilibrium
  conditions
    for chemically reacting systems, 37–39, 82
    for noninteracting systems, 34–37
  effect of pressure, 39
  equation for, 25
  phenomenological theory of, 25
  serum albumin–methyl orange, 83–85
Sedimentation pattern
  classical interpretation of, 23–25
  definition of, 20
  interpretation
    for fast interaction, 103
    for slow interaction, 103
  multiple zones and peaks due to interaction, 93–94
  radial dilution, 23
  shape and area of boundaries, 23
Serum albumin
  interaction
    with amines, 70
    with borate, 189, 192–195, 231
    with Cd^{2+}, 58, 69
    with chloride ion, 57–59, 69
    with fatty acid, 62
    with methyl orange, 57, 62
      electrophoresis, 66–69
      sedimentation equilibrium, 83–85
      velocity sedimentation, 79–81
    with pepsin, 58, 88–91, 93, 146–147
  nonenantiographic electrophoretic patterns of, 54–55
Similarity transformation
  classes of, 96
  definition of, 95
  Gilbert–Jenkins theory, 142
  Gilbert theory, 104
  ideal transport, 101–102
  one-parameter groups of, 96–97
  solution of Fick's second law, 98–100
Similarity variable, 96
Stability, numerical, 210, 212
Stationary electrophoretic boundary, significance of, 57
Stokes law, 5
Svedberg equation, 21–22, 34

T

Tests for heterogeneity, 187, 190, 193–195
  for interactions, 102–103, 109–110, 114–115, 122–123, 144, 157, 179, 187, 190, 193–195
  for polymerization, 226–227
  for pressure effects, 164–166, 170
Total chemical potential, 35, 43–44
Trailing zones due to interactions, 191
Transport equation
  comparison of analytical and numerical treatment, 155
  exact, 152
  ligand-mediated association–dissociation, 196–204
  numerical solution of
    methods, 182, 207–223
    results, 152–206
Transport-interconversion equation, 47
  for ampholytes, 47
    electrophoresis of, 51
  isomerization
    approximate, 94
    exact, 157
  macromolecular isomerization, 48
  polymerization, exact, 155
  analytical solution, isomerization, 94
  approximate, 94–95
  macromolecular complexes, 141
  macromolecule–neutral-molecule interaction, 180
  rationale for ignoring diffusion, 95
Truncation errors, 212, 214
Tryptophan synthetase, chromatography, 175, 233

U

Ultracentrifugation
  analytical, 19–40
  antigen–antibody complexes, 134–136
  in combination with electrophoresis, 10
  definition of, 19
  differential and comparative techniques, 72
  Johnston–Ogston effect, 23, 32, 34

Subject Index

Ultracentrifugation—contd.
  photoelectric scanner, 72–73, 75, 78–79
  preparative, 19
  principles of, 19–43
  radial dilution, 23

V

Velocity sedimentation
  adjustment of pressure, 164
  comparative, 73
  convection, 168, 170
  effect of pressure, 81, 160–171
  fractionation, 225
  interactions, theory of pressure effect, 167–170
  kinetically controlled isomerization, 159–160
  method of, 21
  phenomenological theory of, 25–34
  procedures and precautions, 225–228
  protein–buffer interactions, 195–196
  protein–small-ion interaction, 76–81
  rotor speed dependence, causes, 197
  serum albumin–methyl orange, 79–81
  tests for heterogeneity, 225–226
    for pressure effects, 164–166, 170
Viscosity, 10

W

Weak-electrolyte moving-boundary theory, 50–58, 93, 177
  application
    to certain interacting systems, 47
    to interaction between macromolecules, 58,
    to protein–small-ion interactions, 57–59
  interactions between different macromolecules, 85
  protein–small-ion interaction, 63–64
Wilson's discontinuous solution, 105, 119–121

Z

Zeta potential
  definition of, 6
  evaluation of, 7
Zone electrophoresis
  choice of buffer, 231–232
  conalbumin–borate interaction, 195
  definition of, 11
  in density gradient, 18
  departure from ideality, 18
  determination of mobilities and iso-electric points, 18
  electroosmosis, 18
  factors governing rate of migration, 18
  fractionation, 230–231
  in gels, molecular sieving, 18
  ideal, 17–18
  kinetically controlled isomerization, 159
  macromolecular association–dissociation, 171–173
  macromolecular complexes, 173–175
  multiple zones due to interaction, 230–232
  precautions, 230
  protein–buffer interaction, theory, 190–192
  serum albumin–borate interaction, 192–195
  trailing zones, 191
  two zones from single macromolecule, 171, 192
Zone electrophoretic pattern, concentration dependence, 231
Zone sedimentation
  interacting systems, 175–176
  kinetically controlled isomerization, 159–160
  multiple zoning, 228
  protein–buffer interactions, 195–196
Zone transport, interacting systems, 171–176
Zone-velocity sedimentation in preformed density gradient, principle of, 42